P9-AGR-278

Q.98

An Introduction to ATM Networks

An Introduction to ATM Networks

Harry G. Perros
NC State University, Raleigh, USA

JOHN WILEY & SONS, LTD
Chichester • New York • Weinheim • Brisbane • Singapore • Toronto

Baffins Lane, Chichester,
West Sussex PO19 1UD, England

National 01243 779777
International (+44) 1243 779777

e-mail (for orders and customer service enquiries): cs-books@wiley.co.uk

Visit our Home Page on http://www.wiley.co.uk or http://www.wiley.com

Other Wiley Editorial Offices

John Wiley & Sons, Inc., 605 Third Avenue,
New York, NY 10158-0012, USA

Wiley-VCH Verlag GmbH
Pappelallee 3, D-69469 Weinheim, Germany

John Wiley & Sons Australia, Ltd
33 Park Road, Milton, Queensland 4064, Australia

John Wiley & Sons (Asia) Pte Ltd, 2 Clementi Loop #02-01,
Jin Xing Distripark, Singapore 129809

John Wiley & Sons (Canada) Ltd, 22 Worcester Road,
Rexdale, Ontario, M9W 1L1, Canada

Library of Congress Cataloging-in-Publication Data

Perros, Harry G.
An introduction to ATM networks / Harry G. Perros.
p. cm.
Includes bibliographical references and index.
ISBN 0-471-49827-0 (alk.paper)
1. Asynchronous transfer mode. I. Title.

TK5105.35.P48 2001
004.6′6—dc21

2001026646

British Library Cataloguing in Publication Data

A catalogue record for this book is available from the British Library

ISBN 0-471-49827-0

Typeset in 10/12pt Times Roman by Laser Words, Chennai, India.
Printed and bound in Great Britain by Biddles Ltd, Guildford Surrey.
This book is printed on acid-free paper responsibly manufactured from sustainable forestry in which at least two trees are planted for each one used for paper production.

To

Helen, Nick and Mikey

About the Author

Harry G. Perros received the BSc degree in mathematics in 1970 from Athens University, Greece, the MSc degree in operational research with computing from Leeds University, England, in 1971, and the PhD degree in operations research from Trinity College Dublin, Ireland, in 1975.

From 1976 to 1982 he was an Assistant Professor in the Department of Quantitative Methods, University of Illinois at Chicago. In 1982 he joined the Department of Computer Science, North Carolina State University, as an Associate Professor, and since 1988 he has been a Professor. He has spent sabbaticals at INRIA, Rocquencourt, France, University of Paris 6, France, and NORTEL, Research Triangle Park, North Carolina.

He has published extensively in the area of performance modeling of computer and communication systems, and has organized several national and international conferences. He has also published a monograph entitled *Queueing Networks with Blocking*: *Exact and Approximate Solutions* (Oxford University Press). He is the chairman of the IFIP Working Group 6.3 on the Performance of Communication Systems. In his free time, he likes to sail on board the *Aegean*, a Pearson 31!

Contents

Preface

ATM networks was the subject of intense research and development from the late 1980s to the late 1990s. Currently, ATM is a mature networking technology and is regularly taught in universities and in short professional courses. This book was written with a view to be used as a textbook in a second course on computer networks at the graduate level or senior undergraduate level. Also, it was written for networking engineers out in the field who would like to learn more about ATM networks. A prerequisite for this book is basic knowledge of computer networking principles.

The book is organized into the following parts:

Part One: Introduction and Background
Part Two: The ATM Architecture
Part Three: Deployment of ATM
Part Four: Signaling in ATM Networks.

Part One, 'Introduction and Background', contains a variety of topics which are part of the background necessary for understanding the material in this book. It consists of Chapters 1, 2 and 3. Chapter 1 contains a discussion of what caused the development of ATM networks, and a brief description of the various standards committees that feature prominently in the development of ATM networks. Chapter 2 gives a review of basic concepts of computer networks that are used in this book. This chapter can be skipped by the knowledgeable reader. Chapter 3 is dedicated to frame relay, where we describe the motivation behind the development of frame relay and its basic features, the frame relay UNI, and congestion control. It is educationally constructive to understand how frame relay works, since it is a very popular networking solution and it has many common features with ATM networks, such as layer two switching, no error or flow control between two adjacent nodes, and similar congestion control schemes.

Part Two, 'The ATM Architecture', focuses on the main components of the ATM architecture. It consists of Chapters 4, 5, 6 and 7. In Chapter 4, the main features of the ATM architecture are presented. An ATM packet, known as a *cell*, has a fixed size and it is equal to 53 bytes. We start with a brief account of the considerations that led to the decision to use such a small packet. Then, we describe the structure of the header of the ATM cell, the ATM protocol stack, and the various ATM interfaces. We conclude this chapter with a description of the physical layer that supports ATM networks, and the various public and private interfaces. In Chapter 5, we describe the ATM adaptation layer. The purpose of this layer is to isolate higher protocol layers and applications from the specific characteristics of ATM. Four different ATM adaptation layers are described, namely ATM adaptation layers 1, 2, 3/4 and 5. Chapter 6 is dedicated to ATM switch architectures, and the following different classes of architecture are presented: space-division switches, shared memory switches, and shared medium switches. We describe various architectures that have been proposed within each of these three classes. Also, to give the reader a feel of a real-life switch, the architecture of a commercial switch is described. We conclude this chapter by describing various algorithms for scheduling the transmission of cells out of an output port of an ATM switch. Finally, Chapter 7 deals with the interesting problem of congestion control in ATM networks. We first present the various parameters used to characterize ATM traffic, the various Quality of Service (QoS) parameters, and the standardized ATM classes. In the rest of the chapter, we focus on the two classes of congestion control schemes, namely, preventive and reactive congestion control. We introduce the preventive congestion control scheme, and present various call admission control algorithms, the GCRA bandwidth enforcement algorithm, and cell discard policies. Finally, we present the Available Bit Rate (ABR) scheme, a reactive congestion control scheme standardized by the ATM Forum.

Part Three, 'Deployment of ATM', deals with the different topics: how IP traffic is transported over ATM, and ADSL-based access networks. In Chapter 8, we describe various schemes used to transport IP traffic over ATM. We first present ATM Forum's LAN Emulation (LE), a solution that enables existing LAN applications to run over an ATM network. Then, we describe the IETF's classical IP and ARP over ATM and Next Hop Resolution Protocol (NHRP) schemes, designed for carrying IP packets over ATM. The rest of the chapter is dedicated to three techniques, IP switching, tag switching, and Multi-Protocol Label Switching (MPLS). IP switching inspired the development of tag switching, which at the moment is being standardized by IETF under the name of multi-protocol label switching. Chapter 9 is dedicated to Asymmetric Digital Subscriber Line (ADSL) technology, which can be used in residential access networks to provide basic telephone services and access to the Internet. We describe the Discrete Multi-Tone (DMT) technique used to transmit the information over the telephone twisted pair, the seven bearer channels, the fast and interleaved paths, and the ADSL super frame. Finally, we discuss architectures for accessing network service providers.

Part Four, 'Signaling in ATM Networks', focuses on the signaling protocols used to set-up a Switched Virtual Connection (SVC). In Chapter 10, we review the signaling protocols used to establish a point-to-point connection and a point-to-multipoint connection over the private UNI. The signaling protocol for establishing a point-to-point connection is described in ITU-T's Q.2931 standard, and the signaling protocol for establishing a point-to-multipoint connection is described in ITU-T's Q.2971 standard. We first describe a specialized ATM adaptation layer, known as the signaling AAL (SAAL), which is used by both protocols. Then, we discuss in detail the signaling messages and procedures used by Q.2931 and Q.2971. In Chapter 11, we examine the Private Network-Network Interface (PNNI) used to route a new call from an originating UNI to a destination UNI. PNNI consists of the PNNI routing protocol and the PNNI signaling protocol. We first describe the PNNI routing protocol in detail, and then we briefly discuss the PNNI signaling protocol.

At the end of each chapter there are some problems given. Also, in Chapters 6 and 7 there are three simulation projects, designed to help the reader understand better some of the intricacies of ATM networks.

To develop a deeper understanding of ATM networks, one has to dig into the various documents produced by the standards bodies. Most of these documents are actually very readable! A list of standards which are relevant to the material presented here can be found at the end of the book.

Finally, in the ATM networks field there is an abundance of abbreviations, and the reader is strongly encouraged to learn some of them. When in doubt, the list of abbreviations given may be of help!

Harry Perros

List of Abbreviations

AAL	ATM adaptation layer
ABR	available bit rate
ABT	ATM block transfer
ACR	allowable cell rate
ADSL	asymmetric digital subscriber line
AFI	authority and format identifier
ANP	AAL 2 negotiation procedure
APON	ATM passive optical networks
ARP	address resolution protocol
ARQ	automatic repeat request
ATM	asynchronous transfer mode
ATU-C	ADSL transceiver unit at the central office
ATU-R	ADSL transceiver unit at the remote terminal
BAS	broadband access server
BCOB-A	broadband connection oriented bearer class A
BCOB-C	broadband connection oriented bearer class C
BCOB-X	broadband connection oriented bearer class X
B-frame	bi-directional-coded frame
B-ICI	broadband inter-carrier interface
BECN	backward explicit congestion notification
BGP	border gateway protocol
BOM	beginning of message
BT	burst tolerance
BUS	broadcast and unknown server
CAC	call admission control
CBR	constant bit rate
CCITT	International Telegraph and Telephone Consultative Committee
CCR	current cell rate
CDVT	cell delay variation tolerance
CER	cell error rate
CI	connection identifier
CIDR	classless inter-domain routing
CIR	committed information rate
CLEC	competitive local exchange carrier
CLLM	consolidated link layer management

CLNAP	connectionless network access protocol
CLNIP	connectionless network interface protocol
CLP	cell loss priority bit
CLR	cell loss rate
CLS	connectionless server
CMR	cell misintertion rate
CO	central office
COM	continuation of message
CoS	class of service
CPS	common part sublayer
CRC	cyclic redundant check
CR-LDP	constraint routing-label distribution protocol
CS	convergence sublayer
CTD	cell transfer delay
DBR	deterministic bit rate
DCC	data country code
DCE	data communication equipment
DMCR	desirable minimum cell rate
DMT	discrete multi-tone
DOCSIS	data-over-cable service interim specification
DSL	digital subscriber loop
DSLAM	ADSL access multiplexer
DSP	domain-specific part
DTE	data terminal equipment
DTL	designated transit list
EFCN	explicit forward congestion notification
EOM	end of message
ER	explicit rate
ESI	end system identifier
FCS	frame check sequence
FDM	frequency division multiplexing
FEC	forwarding equivalent class
FECN	forward explicit congestion notification
FIB	forwarding information base
FRAD	frame relay access devices
FRP/DT	fast reservation protocol with delayed transmission
FTTB	fiber to the basement
FTTC	fiber to the curb
FTTCab	fiber to the cabinet
FTTH	fiber to the home
GCRA	generic cell rate algorithm
GFR	guaranteed frame rate
GSMP	general switch management protocol
HDLC	high-level data link control
HDSL	high data rate DSL
HEC	header error control
HFC	hybrid fiber coaxial

HO-DSP	high-order DSP
IBP	interrupted Bernoulli process
ICD	international code designator
ICMP	internet control message protocol
IDI	initial domain identifier
IDP	initial domain part
IDSL	ISDN DSL
IE	information elements
IFP	interrupted fluid process
IFMP	Ipsilon's flow management protocol
I-frame	intra-coded frame
IGMP	internet group management protocol
IISP	interim interswitch signaling protocol
InATMARP	inverse ATMARP
ILEC	incumbent local exchange carrier
IP	internet protocol
IPP	interrupted Poisson process
ISO	International Organization of Standards
ISP	Internet service provider
ITU	International Telecommunication Union
IWU	interworking unit
L2TP	layer 2 tunnel protocol
LAC	L2TP access concentrator
LDP	label distribution protocol
LE	LAN emulation
LE-ARP	LAN emulation address resolution
LECID	LE client identifier
LER	label edge router
LIS	logical IP subnet
LIJ	leaf initiated join
LMDS	local multipoint distribution services
LMI	local management interface
LSP	label switched path
LSR	label switching router
LUNI	LAN emulation user to network interface
MARS	multicast address resolution server
MBS	maximum burst size
MCR	minimum cell rate
MCS	multicast servers
ME	mapping entity
MFS	maximum frame size
MMBP	Markov modulated Bernoulli process
MMPP	Markov modulated Poisson process
MPLS	multi-protocol label switching
MPOA	multi-protocol over ATM
MTU	maximum transfer unit
NAS	network access server

NBMA	non broadcast multiaccess network
NHC	next hop client
NHRP	next hop resolution protocol
NHS	next hop server
NNI	network node interface
NRT-VBR	non-real-time variable bit rate
NRT-SBR	non-real-time statistical bit rate
NSAP	network service access point
NSP	network service provider
NTR	network timing reference
OC	optical carrier
OLT	optical line terminator
ONU	optical network unit
OSI	open system interconnection reference model
OSPF	open shortest path first
PCM	pulse code modulation
PCR	peak cell rate
PDH	plesiochronous digital hierarchy
PDU	protocol data unit
P-frame	predictive-coded frame
PGL	peer group leader
PIM	protocol independent multicast
PMD	physical medium dependent sublayer
PNNI	private network-network interface or private network node interface
PON	passive optical network
PPP	point-to-point protocol
PTI	payload type Indicator
PTSE	PNNI topology state element
PTSP	PNNI topology state packet
PVC	permanent virtual connection
QAM	quadrature amplitude modulation
RADIUS	remote authentication dial in user service
RCC	routing control channel
RM	resource management
ROC	regional operations center
RSVP	resource reservation protocol
RT-VBR	real-time variable bit rate
RT-SBR	real-time statistical bit rate
SAAL	signaling AAL
SAR	segmentation-and-reassembly sublayer
SBR	statistical bit rate
SCR	sustained cell rate
SDH	synchronous digital hierarchy
SDU	service data unit
SDSL	symmetric DSL
SEL	selector
SMDS	switched multimegabit data service

SONET	synchronous optical network
SSCF	service-specific connection function
SSCOP	service-specific connection oriented protocol
SSCS	service specific convergence sublayer
SSM	single segment message
STF	start field
STM	synchronous transfer mode
STS-1	synchronous transport signal level 1
SVC	switched virtual connection
TC	transmission convergence sublayer
TDP	tag distribution protocol
TER	tag edge router
TFIB	tag forwarding information base
TSR	tag switching router
TTL	time to live
UBR	unspecified bit rate
UNI	user network interface
VCC	virtual channel connection
VCI	virtual channel identifier
VDSL	very high data rate DSL
VPI	virtual path identifier
WDM	wavelength division multiplexing
xDSL	x-type digital subscriber line

Part 1

Introduction and Background

1

Introduction

In this chapter, we introduce the Asynchronous Transfer Mode (ATM) networking technique, and discuss the forces that gave rise to it. Then, we describe some of the well known national and international standards committees involved with the standardization process of networking equipment.

1.1 THE ASYNCHRONOUS TRANSFER MODE

ATM is a technology that provides a single platform for the transmission of voice, video and data at specified quality of service and at speeds varying from fractional T1 (i.e. nX64 Kbps), to Gbps. Voice, data and video are currently transported by different networks. Voice is transported by the public telephone network, and data by a variety of packet-switched networks. Video is transported by networks based on coaxial cables, satellites and radio waves, and to a limited extent, by packet-switched networks.

To understand what caused the development of ATM, we have to go back to the 1980s! During that decade, we witnessed the development of the workstation and the evolution of the optical fiber. A dramatic reduction in the cost of processing power and associated peripherals, such as main memory and disk drives, led to the development of powerful workstations capable of running large software. This was a significant improvement over the older 'dumb terminal'. These workstations were relatively cheap to buy, easy to install and interconnect, and they enabled the development of distributed systems. As distributed systems became more commonplace, so did the desire to move files over the network at a higher rate. Also, there was a growing demand for other applications, such as videoconferencing, multimedia, medical imaging, remote processing and remote printing of a newspaper. At the same time, optical fiber technology evolved very rapidly, and by the end of the 1980s a lot of optical fiber had been installed. Optical fiber permitted high bandwidth and very low bit-error rate.

These technological developments, coupled with the market needs for faster interconnectivity, gave rise to various high-speed wide-area networks and services, such as *frame relay*, *Asynchronous Transfer Mode* (ATM) and *Switched Multimegabit Data Services* (SMDS).

ATM was standardized by ITU-T in 1987. It is based on packet-switching and is connection oriented. An ATM packet, known as a *cell*, is a small fixed-size packet with a payload of 48 bytes and a 5-byte header. The reason for using small packets was motivated mostly by arguments related to the transfer of voice over ATM.

Unlike IP networks, ATM has built-in mechanisms that permit it to provide different quality of service to different types of traffic. ATM was originally defined to run over high-speed links. For instance, in North America, the lowest envisioned speed was OC-3, which corresponds to about 155 Mbps. It should be noted that the fastest network in the late 1980s was the FDDI (Fiber Distributed Data Interface), which ran at 100 Mbps. However, as ATM became more widely accepted, it was also defined over slow links, such as fractional T1, i.e., nX64 Kbps.

In the early 1990s, ATM was poised to replace well-established local and wide area networks such as Ethernet and IP networks. ATM was seen as a potential replacement for Ethernet because it ran faster, and also provided a good quality of service. At that time, Ethernet ran at 10 Mbps, but due to software bottlenecks, its effective throughput was around 2 Mbps. Also, since ATM has its own addressing system, and it can set-up and route connections through the network, it was seen as a potential foe of IP networks. In view of this, Ethernet and IP networks were declared by the ATM aficionados as 'dead'!

Interestingly enough, Ethernet made a dramatic come-back when it was defined to run at 100 Mbps and later on at 1 Gbps. As a result, ATM lost the battle to the 'desktop', i.e. it never became the preferred networking solution for interconnecting workstations and personal computers at a customer's premises. Also, in the mid-1990s, we witnessed a new wave of high-speed IP routers and a strong effort to introduce quality of service in IP networks. As a result, one frequently hears cries that it is the ATM technology that is now 'dead'!

ATM is a mature networking technology, and it is still the only networking technology that provides quality of service. ATM networks are used in a variety of environments. For instance, it is widely used in the backbone of Internet Service Providers (ISP) and in campus networks to carry Internet traffic. ATM wide area networks have also been deployed to provide point-to-point and point-to-multipoint video connections. Also, there are on-going projects in telecommunication companies aiming at replacing the existing trunks used in the telephone network with an ATM network.

On a smaller scale, ATM is used to provide *circuit emulation*, a service that emulates a point-to-point T1/E1 circuit and a point-to-point fractional T1/E1 circuit over an ATM network. ATM is the preferred solution for ADSL-based residential access networks used to provide access to the Internet and basic telephone services over the phone line. Also, it is used in Passive Optical Networks (PON) deployed in residential access networks.

We conclude this section by noting that arguments in favor and against existing and emerging new networking technologies will most likely continue for a long time. There is no argument, however, that these are indeed very exciting times as far as communication systems are concerned!

1.2 STANDARDS COMMITTEES

Standards allow vendors to develop equipment to a common set of specifications. Providers and end-users can also influence the standards so that the vendors' equipment conforms to certain characteristics. As a result of the standardization process, one can purchase equipment from different vendors without being bound to the offerings of a single vendor.

There are two types of standards, namely *de facto* and *de jure*. *De facto* standards are those which were first developed by a single vendor or a consortium, and then they were accepted by the standards bodies. *De jure* standards are those generated through consensus within national or international standards bodies. ATM, for instance, is the result of the latter type of standardization.

Several national and international standards bodies are involved with the standardization process in telecommunication, such as the International Telecommunication Union (ITU), the International Organization for Standardization (ISO), the American National Standards Institute (ANSI), the Institute of Electrical and Electronics Engineering (IEEE), the Internet Engineering Task Force (IETF), the ATM Forum, and the Frame Relay Forum. The organizational structure of these standards bodies is described below.

The ITU-T and the ATM Forum are primarily responsible for the development of standards for ATM networks. ITU-T concentrates mainly on the development of standards for public ATM networks, whereas the ATM Forum concentrates on private networks. The ATM Forum was created because many vendors felt that the ITU-T standardization process was not moving fast enough, and also because there was an emerging need for standards for private ATM networks. In general, ITU-T tends to reflect the view of network operators and national administrations, whereas the ATM Forum tends to represent the users and the Customer Premise Equipment (CPE) manufacturers. The two bodies compliment each other and work together to align their standards with each other.

The International Telecommunication Union (ITU)

ITU is a United Nations specialized agency whose job is to standardize international telecommunications. ITU consists of the following three main sections: the ITU Radiocommunications Sector (ITU-R), the ITU Telecommunications Standardization Sector (ITU-T), and the ITU Development Sector (ITU-D).

The ITU-T's objective is telecommunications standardization on a worldwide basis. This is achieved by studying technical, operating and traffic questions, and adopting recommendations on them. ITU-T was created in March 1993, and it replaced the former well-known standards committee, the International Telegraph and Telephone Consultative Committee, whose origins go back over 100 years. This committee was commonly referred to as the CCITT, which are the initials of its name in French.

ITU-T is formed by representatives from standards organizations, service providers, and more recently, by representatives from vendors and end users. Contributions to standards are generated by companies, and they are first submitted to national technical coordination groups, resulting in national standards. These national coordinating bodies may also pass on contributions to regional organizations, or directly to ITU-T, resulting in regional or world standards. ITU more recently started recommending and referencing standards adopted by the other groups, instead of rewriting them.

ITU-T is organized into 15 technical study groups. At present, more than 2500 recommendations (standards) or some 55 000 pages are in force. They are nonbinding standards agreed by consensus in the technical study groups. Although, nonbinding, they are generally complied with due to their high quality, and also because they guarantee the interconnectivity of networks, and enable telecommunications services to be provided on a worldwide scale.

ITU-T standards are published as *recommendations*, and they are organized into series. Each series of recommendations is referred to by a letter of the alphabet. Some of the well-known recommendations are the I, Q and X. Recommendations I are related to integrated services digital networks. For instance, I.321 describes the B-ISDN protocol reference architecture, I.370 deals with congestion management in frame relay, and I.371 deals with congestion management in ATM networks. Recommendations Q are related to switching and signaling. For instance, Q.2931 describes the signaling procedures used to establish a point-to-point ATM switched virtual connection over the private UNI, and Q.2971 describes the signaling procedures used to establish a point-to-multipoint ATM switched virtual connection over the private UNI. Recommendations X are related to data networks and open system communication. For instance, X.700 describes the management framework for the OSI basic reference model, and X.25 deals with the interface between a DTE and a DCE terminal operating in a packet mode and connected to a public data network by a dedicated circuit.

The International Organization for Standardization (ISO)

ISO is a worldwide federation of national standards bodies from some 130 countries, one from each country. It is a nongovernmental organization established in 1947. Its mission is to promote the development of standardization and related activities in the world, with a view to facilitating the international exchange of goods and services, and to developing cooperation in the spheres of intellectual, scientific, technological and economic activity.

It is interesting to note that the name ISO does not stand for the initials of the full title of this organization, which would have been IOS! In fact, ISO is a word derived from the Greek *isos*, which means 'equal'. From 'equal' to 'standard' was the line of thinking that led to the choice of ISO. In addition, the name ISO is used around the world to denote the organization, thus avoiding a plethora of acronyms resulting from the translation of 'International Organization for Standards' into the different national languages of the ISO members, such as IOS in English, and OIN in French (from Organization International de Normalization).

ISO's standards covers all technical fields. Well known examples of ISO standards are: the ISO film speed code, the standardized format of telephone and banking cards, ISO 9000 which provides a framework for quality management and quality assurance, paper sizes, safety wire ropes, ISO metric screw threads, and the ISO international codes for country names, currencies and languages. In telecommunications, the Open System Interconnection (OSI) reference model (see Chapter 2) is a well known ISO standard.

ISO has co-operated with the International Electronical Commission (IEC) to develop standards in computer networks. IEC emphasizes hardware, while ISO emphasizes software. In 1987 the two groups formed the Joint Technical Committee 1 (JTC 1). This committee developed documents that became ISO and IEC standards in the area of information technology.

The American National Standards Institute (ANSI)

ANSI is a nongovernmental organization formed in 1918 to act as a cross between a standards setting body and a coordinating body for US organizations that develop standards. ANSI represents the US in international standards bodies such as ITU-T and ISO. ANSI is not restricted to information technology. In 1960 ANSI formed X3, a committee responsible for developing standards within the information processing area in the US. X3 is made up of 25 technical committees, of which X3S3 is the committee responsible for data communications. The main telecommunications standards organization within ANSI is the T1 secretariat, sponsored by the Exchange Carriers Standards Association. ANSI is focused on standards above the physical layer. Hardware oriented standards are the work of the Electronics Industries Association (EIA) in the US.

The Institute of Electrical and Electronics Engineering (IEEE)

IEEE is the largest technical professional society in the world, and it has been active in developing standards in the area of electrical engineering and computing through its IEEE Standards Association (IEEE-SA). This is an international organization with a complete portfolio of standards. The IEEE-SA has two governing bodies: the Board of Governors, and the Standards Board. The Board of Governors is responsible for the policy, financial oversight, and strategic direction of the Association. The Standards Board has the charge to implement and manage the standards process, such as approving projects.

One of the most well known IEEE standards bodies in the networking community is the LAN/MAN Standards Committee, or otherwise known as the IEEE project 802. They are responsible for several well known standards, such as CSMA/CD, token bus, token ring, and the Logical Link Control (LLC) layer.

The Internet Engineering Task Force (IETF)

The IETF is part of a hierarchical structure that consists of the following four groups: the Internet Society (ISOC) and its Board of Trustees, the Internet Architecture Board (IAB), the Internet Engineering Steering Group (IESG), and the Internet Engineering Task Force (IETF) itself.

The ISOC is a professional society concerned with the growth and evolution of the Internet worldwide. The IAB is a technical advisory group of the ISOC, and its charter is to provide oversight of the Internet and its protocols, and to resolve appeals regarding the decisions of the IESG. The IESG is responsible for technical management of IETF activities and the Internet standards process. It administers the standardization process according to the rules and procedures which have been ratified by the ISOC Trustees.

The IETF is a large open international community of network designers, operators, vendors and researchers concerned with the evolution of the Internet architecture and the smooth operation of the Internet. It is divided into the following eight functional areas: applications, Internet, IP: next generation, network management, operational requirements, routing, security, transport, and user services. Each area has several working groups. A working group is made up of a group of people who work under a charter in order to achieve a certain goal. Most working groups have a finite lifetime, and a working group is dissolved once it has achieved its goal. Each of the eight functional areas has one or two area directors, who are members of IESG. Much of the work of IETF is handled via mailing lists, which anyone can join.

The IETF standards are known as Request For Comments (RFC), and each of them is associated with a different number. For instance, RFC 791 describes the Internet Protocol (IP), and RFC 793 the Transmission Control Protocol (TCP). Originally, an RFC was just what the name implies, that is, a request for comments. Early RFCs were messages between the ARPANET architects about how to resolve certain procedures. Over the years, however, RFCs became more formal, and they were cited as standards, even when they were not. There are two subseries within the RFCs, namely, For Your Information (FYI) RFCs and standard (STD) RFCs. The FYI RFC subseries was created to document overviews and topics which are introductory in nature. The STD RFC subseries was created to identify those RFCs which are in fact Internet standards.

Another type of Internet document is the *Internet-draft*. These are work-in progress documents of the IETF, submitted by any group or individual. These documents are valid for six months, and they may be updated, replaced, or they may become obsolete.

Finally, we note that the ISOC has also chartered the Internet Assigned Numbers Authority (IANA) as the central coordinator for the assignment of 'unique parameters' on the Internet, including IP addresses.

The ATM Forum

During the late 1980s, many vendors felt that the ATM standardization process in ITU-T was too slow. The ATM Forum was created in 1991 with the objective of accelerating the use of ATM products and services in the private domain through a rapid development of specifications. The ATM Forum is an international, nonprofit organization, and it has generated very strong interest within the communications industry. Currently, it consists of over 600 member companies, and it remains open to any organization that is interested in accelerating the availability of ATM-based solutions.

The ATM Forum consists of the *Technical Committee*, three *Market Awareness Committees* for North America, Europe and Asia-Pacific, and the *User Committee*.

The ATM Forum Technical Committee works with other worldwide standards bodies selecting appropriate standards, resolving differences among standards, and recommending new standards when existing ones are absent or inappropriate. It was created as a single worldwide committee in order to promote a single set of specifications for ATM products and services. It consists of several working groups, which investigate different areas of ATM technology, such as the ATM architecture, routing and addressing, traffic management, ATM/IP collaboration, voice and multimedia over ATM, control signaling, frame-based ATM, network management, physical layer, security, wireless ATM, and testing.

The ATM Market Awareness Committees provide marketing and educational services designed to speed the understanding and acceptance of ATM technology. They coordinate the development of educational presentation modules and technology papers, publish the *53 Bytes*, the ATM Forum's newsletter, and coordinate demonstrations of ATM at trade shows.

The ATM Forum User Committee, formed in 1993, consists of organizations which focus on planning, implementation, management or operational use of ATM-based networks, and network applications. This committee interacts regularly with the Market Awareness Committees and the Technical Committee to ensure that ATM technical specifications meet real-world end-user needs.

The Frame Relay Forum

The Frame Relay Forum was formed in 1991, and is an association of vendors, carriers, users and consultants committed to the implementation of frame relay in accordance with national and international standards.

The Forum's technical committees take existing standards, which may not be sufficient for full interoperability, and create Implementation Agreements (IA). These IAs represent an agreement by all members of the frame relay community as to the specific manner in which standards will be applied. At the same time, the Forum's marketing committees are chartered with worldwide market development through education as to the benefits if frame relay.

PROBLEMS

1. Visit the web sites of ITU-T, the ATM Forum and IETF. Familiarize yourself with their organizational structure, and the type of standards that are available on these web sites.
2. Read some of the issues of *53 Bytes*, the ATM Forum's newsletter, available on the ATM Forum's web site.

2

Basic Concepts from Computer Networks

In this chapter, we review some basic concepts from computer networks that we use in this book. First, we discuss the various communication networking techniques and the OSI reference model. Then, we present the data link layer of the OSI model, the *High-level Data Link Control* (HDLC), the *synchronous Time Division Multiplexing* (TDM) technique, and the *Logical Link Control* (LLC) layer. Finally, we examine the network access protocol X.25, and conclude with the very popular and important *Internet Protocol version 4* (IPv4).

2.1 COMMUNICATION NETWORKING TECHNIQUES

Communication networking techniques can be classified into the following two broad categories: *switched* and *broadcast* communication networks. Examples of switched communication networks are circuit-switched networks, such as the public telephone system, and packet-switched networks, such as computer networks based on TCP/IP. Examples of broadcast communication networks are packet radio networks, satellite networks, and multi-access local networks such as Ethernet. ATM networks belong to the packet-switched networks.

Circuit switching and packet switching are two different technologies that evolved over a long period of time. Circuit switching involves three phases: *circuit establishment*, *data transfer* and *circuit disconnect*. These three phases take place when we make a phone call. Circuit establishment takes place when we dial up a number. At that moment, the public network attempts to establish a connection to the phone set that we dialed. This involves finding a path to the called party, allocating a channel on each transmission link on the path, and alerting the called party. The data transfer phase follows, during which we converse with the person we called. Finally, the circuit disconnect phase takes place when we hang up. At that moment, the network tears down the connection, and releases the allocated channel on each link on the path. In circuit switching, channel capacity is dedicated for the duration of the connection, even when no data is being sent. For instance, when we make a phone call, the channel that is allocated on each transmission link along the path from our phone to the one we called is not shared with any other phone calls. Also, in circuit switching both stations must be available at the same time in order to establish a connection. Circuit switching is a good solution for voice, since it involves exchanging a relatively continuous flow of data. However, it is not a good solution if the data is bursty. That is, the source emitting the data is active transmitting for a period of time, and then it becomes silent for a period of time during which it is not transmitting. This cycle of being active and then silent repeats until the source completes its transmission. Such an intermittent type of transmission occurs in data transfers. In such cases, the utilization of the circuit-switched connection is low.

Packet switching is appropriate for data exchange. Information is sent in packets, and each packet has a header with the destination address. A packet is passed through the network from node to node until it reaches its destination. Error and flow control procedures can be built into the network to ensure a reliable service. In packet switching, two different techniques can be used, *virtual circuits* and *datagrams*.

A virtual circuit imitates circuit switching, and it involves the same three phases: call set-up, transfer of packets, and call termination. In call set-up, a logical connection is established between the sender and the receiver before any packets are allowed to be sent. This is a path through the nodes of the computer network which all packets will follow. Unlike circuit switching, channel capacity on each transmission link is not dedicated to a virtual circuit. Rather, the transmission link is shared by all the virtual circuits that pass through it. Error control ensures that all packets are delivered correctly in sequence. Flow control is used to ensure that the sender does not over-run the receiver's input buffer. The X.25 network is a good example of a packet-switched network with virtual circuits. Also, as we will see in Chapter 4, ATM networks are also packet-switched networks, and they use virtual circuits.

In datagrams, no call set-up is required, and each packet is routed through the network individually. Because of this, it is possible that two successive packets transmitted from the same sender to the same receiver may follow different routes through the network. Since each packet is routed through the network individually, a datagram service can react to congestion easier. The datagram service provided by the early packet-switched networks was in some cases more primitive than that provided by virtual circuits. For instance, there was no error control, no flow control, and no guarantee of delivering packets in sequence. The IP network, used in the Internet, is a packet-switched network based on datagrams. However, due to the use of static routes in the IP routers, IP packets follow the same path from a sender to a destination, and therefore they are delivered in sequence. Also, unlike earlier packet-switched networks with datagram services, TCP/IP provides both error and flow control.

An example of how two nodes communicate using circuit switching, virtual circuits, and datagrams is given in Figure 2.1. In this example, node 1 communicates with node 4 through intermediate nodes 2 and 3. The passage of time is indicated on the vertical lines, and there is one vertical line per node. In the circuit switching case, the time it takes node 1 to transmit the call request packet and the message is indicated vertically between the two arrows on the first line associated with node 1. The two diagonal parallel lines between the vertical lines of the first and the second nodes show the propagation delay of the call request packet between these two nodes. Similar notation is used for the virtual circuit and datagrams cases. As we can see, the datagram scheme takes less time to transmit the three packets than the virtual circuit scheme.

A broadcast network has a single communication channel that is shared by all the stations. There are no switching nodes as in circuit or packet switching. Data transmitted by one station is received by many, and often by all. An access control technique is used to regulate the order in which stations transmit. The most widespread example of a broadcast network is the Ethernet.

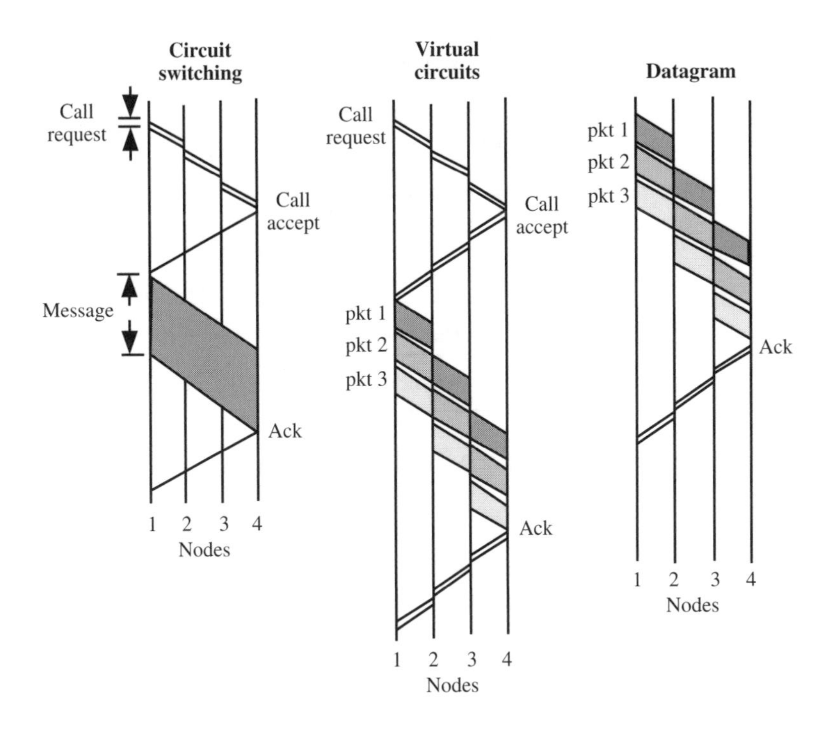

Figure 2.1 A comparison between circuit-switching, virtual circuits and datagrams.

2.2 THE OPEN SYSTEM INTERCONNECTION (OSI) REFERENCE MODEL

In the early days of packet switching, the various communications software suites that were available could not communicate with each other. To standardize the communications protocols, and also facilitate their development, the International Organization for Standardization (ISO) proposed a model known as the *Open Systems Interconnection (OSI) Reference Model*. The functionality of the software for packet switching was grouped into seven layers, namely, the *physical* layer, the *data link* layer, the *network* layer, the *transport* layer, the *session* layer, the *presentation* layer, and the *application* layer. These layers are shown in Figure 2.2. Each layer provides service to the layer directly above it, and receives service from the layer directly below it.

The physical layer is concerned with the transmission of raw bits over a communications channel. The data link's function is to transform the raw transmission link provided by the physical layer into a reliable communications link. This was deemed necessary since early transmission links were inherently unreliable. Modern fiber-based communications links are highly reliable, and as will be seen later on in this book, there is no need for all the data link functionality. The network layer is concerned with routing packets from source to destination, congestion control, and internetworking. The transport protocol is concerned with the end-to-end packet transfer, that is, between an application in the source computer and an application in the destination computer. Some of its main functions are establishment and deletion of connections, reliable transfer of packets, and flow control. The session layer allows users in different computers to set up sessions

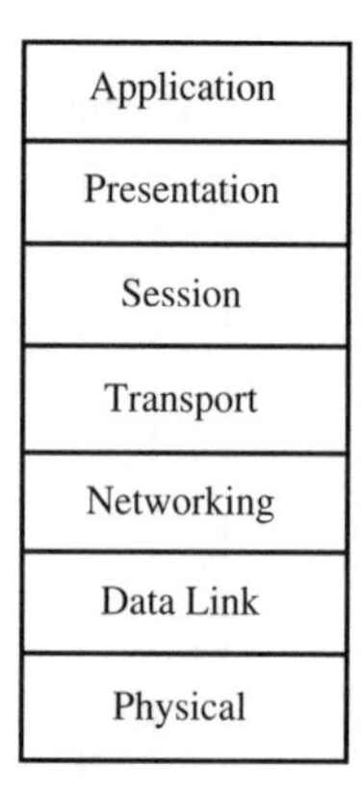

Figure 2.2 The OSI reference model.

between themselves. One of the services of the session layer is to manage dialogue control. The presentation layer is concerned with the syntax and semantics of the information transmitted. In general, two heterogeneous computers may not have the same way of representing data types internally. The presentation layer facilitates the communication between two such computers, by converting the representation used inside a computer to a network standard representation and back. Finally, the application layer contains protocols that are commonly used, such as file transfer, electronic mail and remote job entry.

2.3 DATA LINK LAYER

This protocol layer was designed to provide a reliable point-to-point connection over an unreliable link. The main functions of the data link layer are: window flow control, error control, frame synchronization, sequencing, addressing, and link management. At this layer, a packet is referred to as a *frame*. Below, we examine the window-flow control mechanism, error detection schemes, and the error control mechanism.

Window-flow control

This is a technique for ensuring that a transmitting station does not over-run the receiving station's buffer. The simplest scheme is *stop-and-wait*. The sender transmits a single frame and then waits until the receiver gets the frame and sends an acknowledgment (ACK). When the sender receives the ACK, it transmits a new frame. This scheme is shown in Figure 2.3. The link's utilization U depends on the propagation delay, t_{prop}, and on the time to transmit a frame, t_{frame}. Let

$$a = \frac{t_{\text{prop}}}{t_{\text{frame}}}$$

Then,

$$U = \frac{t_{\text{frame}}}{t_{\text{frame}} + 2t_{\text{prop}}} = \frac{1}{1 + 2a}$$

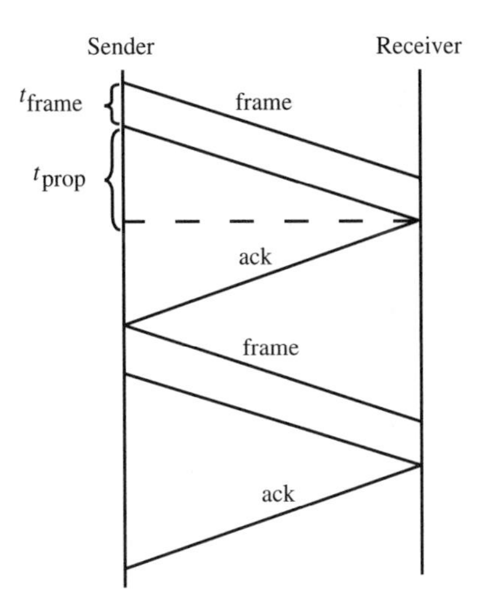

Figure 2.3 The stop-and-wait scheme.

If $a \lll 1$, that is the propagation delay is significantly less than the time to transmit a frame, then the link's utilization U is large. If $a \ggg 1$, that is the propagation delay is significantly greater than the time to transmit a frame, then U is small. As an example, let us consider a satellite link transmitting at 56 Kbps, and let us assume 4000-bit frames and a propagation delay of 270 ms. Then, the time to transmit a frame is 71 ms, $a = 270/71 = 3.8$, and $U = 0.116$.

In the stop-and-wait protocol, only one frame is outstanding (i.e. unacknowledged) at a time. A more efficient protocol is the *sliding window-flow control* protocol, where many frames can be outstanding at a time. The maximum number of frames, W, that a station is allowed to send to another station without acknowledgment is referred to as the *maximum window*. To keep track of which frames have been acknowledged, each frame is numbered sequentially, and the numbers are reusable. An example of the sliding window-flow control scheme is shown in Figure 2.4. The maximum window size W is fixed to 8. In Figure 2.4(a), station A transmits four frames with sequence numbers 1, 2, 3 and 4, and its window is reduced to four, consisting of the sequence numbers $\{5, 6, 7, 8\}$. In Figure 2.4(b), station A sends two more frames with sequence numbers 5 and 6, and its window is down to two, consisting of the numbers $\{7, 8\}$. In Figure 2.4(c), station A receives an ACK from station B for the frames with sequence numbers 1, 2 and 3, and its window opens up to five frames consisting of the sequence numbers $\{7, 8, 1, 2, 3\}$.

The efficiency of this protocol depends upon the maximum window size and the round-trip delay. Let $t_{frame} = 1$. Then,

$$a = \frac{t_{prop}}{t_{frame}} = t_{prop}$$

The time to transmit the first frame and receive an acknowledgment is equal to $t_{frame} + 2t_{prop} = 1 + 2a$. If $W > 1 + 2a$, then the acknowledgment arrives at the sender before the

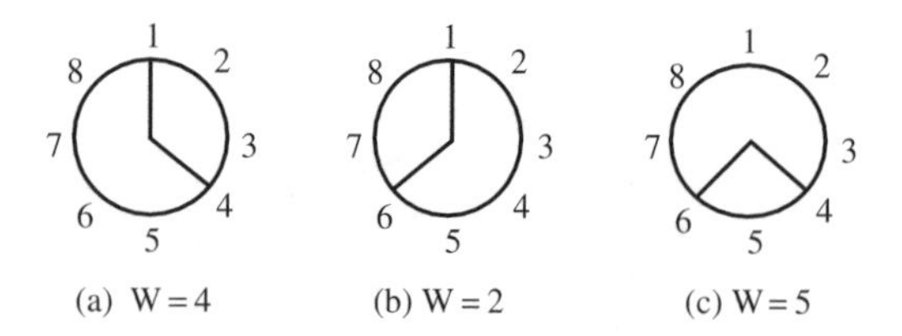

Figure 2.4 An example of the sliding window-flow control scheme.

window has been exhausted, and we have that $U = 1$. If $W < 1 + 2a$, then the acknowledgment arrives after the window has been exhausted, and we have

$$U = \frac{W}{1 + 2a}$$

Error detection

The simplest error detection scheme is the *parity check*. In this scheme, a parity bit is appended to the end of each frame. A more complex error detection scheme based on the parity check is the *longitudinal redundancy check*. The data is organized into a matrix, as shown in Figure 2.5. There are eight columns, and as many rows as the number of bytes. Each matrix element contains one bit. An even parity check is applied to each row and each column. We observe that the parity bit applied to the last column, which contains the parity bits of all the rows, is the same as that applied to the last row which contains the parity bits of all the columns!

The *Cyclic Redundant Check* (CRC) is a commonly used error detection scheme, and is used extensively in ATM networks. The CRC scheme utilizes a predetermined bit pattern P, which is known to both the sender and the receiver. Let $n + 1$ be the length of this bit pattern. Now, let us assume that we have a k-bit message M to be transmitted. The sender shifts M to the left by n bits to obtain the quantity $2^n M$, and then divides $2^n M$ by P. The remainder of that division is an n-bit sequence, known as the *Frame Check Sequence* (FCS). The FCS is added to $2^n M$ and the entire $(k + n)$-bit message is transmitted to the

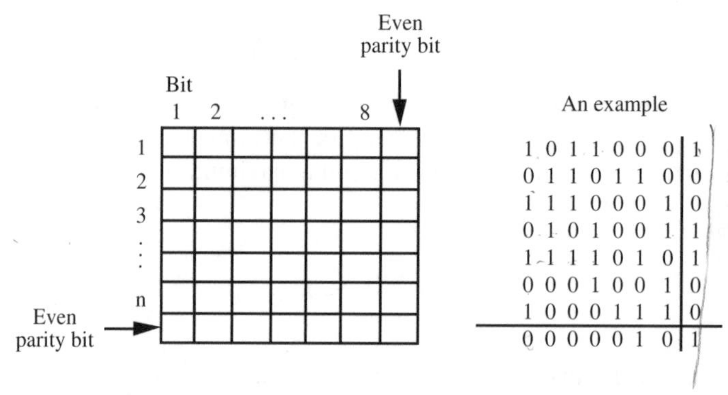

Figure 2.5 The longitudinal redundancy check.

receiver. The receiver divides the message by the same bit pattern P. The message has been received correctly if the remainder of that division is zero. All single bit errors, and some combinations of erroneous bits, can be detected and corrected.

As an example let $M = 1010001101$ and $P = 110101$. Then, the FCS will be five bits long and it is calculated as follows. M is first shifted to the left by five positions, that is $2^5M = 101000110100000$. Then, 2^5M is divided by P, resulting in an FCS equal to 01110. Finally, the transmitted message is 101000110101110. If this message is correctly received, when divided by $P = 110101$, it should give a zero remainder.

It is customary to express the bit pattern P in polynomial form. This is done as follows. Each bit is represented by a term x^n, where n is the location of the bit in the pattern, counting from the right-hand side towards the left-hand side. That is, the rightmost bit corresponds to the term x^0, the second rightmost bit corresponds to the term x^1, and so on. The value of the bit is the coefficient of its corresponding polynomial term. For instance, the pattern 110101 used above is expressed as $x^5 + x^4 + x^2 + 1$.

The *checksum* is another error detection technique that is used in the TCP/IP suite of protocols. The data to be sent is treated as a sequence of binary integers of 16 bits each, and the sum of these 16-bit integers is computed. The data could be of any type or a mixture of types. It is simply treated as a sequence of integers for the purpose of computing their sum. The 16-bit half-words are added up using 1's compliment arithmetic. The 1's compliment of the final result is then computed, which is known as the checksum. 32-bit integers can also be used. The checksum is used in TCP to protect the entire packet, i.e. it is calculated using the header and the payload of the TCP packet. It also used in IP to protect the IP header only. Computing the checksum in TCP is a time-consuming operation, and a considerable speed up can be achieved if it is done in hardware.

Error control

Error control refers to the mechanism used to detect and correct errors that have occurred in the transmission of frames. This mechanism is known as the *Automatic Repeat Request* (ARQ), and it uses error detection, the window-flow control mechanism, positive and negative acknowledgments, and timers. Errors in the transmission of frames occur because a frame is lost or because it is damaged, that is, one or more of its bits have been flipped. Damaged frames are detected by the ARQ mechanism using CRC, and lost frames are detected by observing out-of-sequence frames. Recovery of a lost or damaged frame is done by requesting the sender to re-transmit the frame. Three different versions of the ARQ have been standardized, namely *stop-and-wait ARQ*, *go-back-n ARQ* and *selective-reject ARQ*. The stop-and-wait ARQ is based on the stop-and-wait window-flow control scheme, whereas the go-back-*n* ARQ and the selective-reject ARQ are based on the sliding window-flow control scheme.

In the go-back-*n* scheme, the sender sends a series of frames using the sliding window-flow control technique. Let us assume that station A is transmitting to station B. If B receives a frame correctly, then it sends an ACK with the next frame number that it expects to receive. An ACK may be for several successive frames that have been correctly received. If B receives a damaged frame, say frame i, and it has previously received correctly frame $i - 1$, then B sends a negative acknowledgment (NAK), indicating that frame i is in error. When A receives the NAK, it retransmits frame i plus all other frames after i that it has already transmitted. An example of this scheme is shown in Figure 2.6.

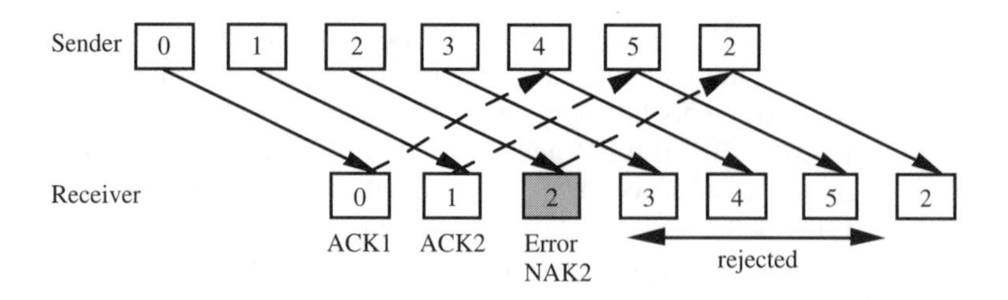

Figure 2.6 The go-back-n scheme.

Now, let us consider the case where frame i is lost. If B correctly receives frame $i + 1$ later on, then it will realize that frame $i + 1$ is out-of-sequence, and it will deduce that frame i is lost. B will then send a NAK, indicating that the ith frame has to be retransmitted. A retransmits frame i plus all other frames after i that it has already transmitted. If frame i is lost and no other frames arrive, then B cannot detect the lost frame. However, for each transmitted frame, A sets a timer. If the timer expires before A receives an ACK or a NAK, A retransmits the frame. In the above case, the lost frame's timer will expire and A will re-transmit it.

In the selective-reject ARQ scheme, only the frame that is in error is retransmitted. All subsequent frames that arrive at B are buffered, until the erroneous frame is received again. This is a more efficient procedure, but it is more complex to implement. The selective-reject scheme is used in TCP. An example of the selective-reject ARQ scheme is shown in Figure 2.7.

2.4 THE HIGH DATA LINK CONTROL (HDLC) PROTOCOL

This protocol has been widely used, and it has been the basis for many other important data link protocols. It was derived from IBM's data link protocol *Synchronous Data Link Control* (SDLC). Later on it was modified and standardized by ISO as the *High Data Link Control* (HDLC) protocol. HDLC was designed to satisfy different types of stations, link configurations and transfer modes. The following three types of stations were defined: *primary*, *secondary* and *combined*. A primary station is responsible for controlling the operation of the link, a secondary station operates under the control of a primary station,

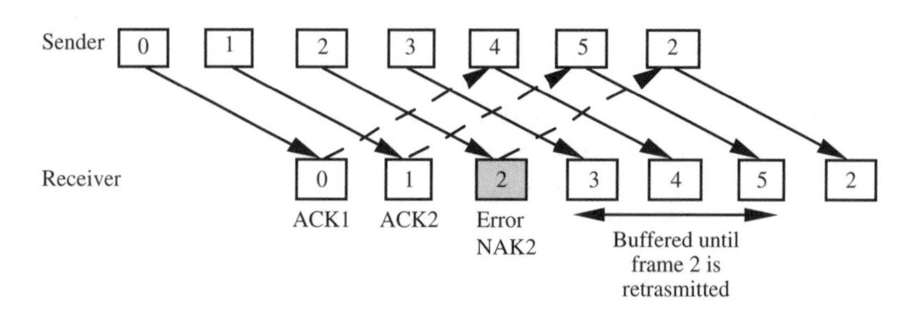

Figure 2.7 The selective-reject scheme.

and a combined station has the features of both the primary and the secondary station. Also, the following types of link configurations were defined: *unbalanced* and *balanced*. An unbalanced configuration consists of one primary and one or more secondary stations, and it supports both full-duplex and half-duplex transmission. A balanced configuration consists of two combined stations, and it supports both full-duplex and half-duplex transmission. Based on these station types and configurations, the following three data transfer modes were defined: *Normal Response time Mode* (NRM), *Asynchronous Balanced Mode* (ABM), and *Asynchronous Response Mode* (ARM). NRM is used with an unbalanced configuration. The primary station initiates data transfers to the secondary stations, and a secondary station may only transmit data in response to a command from the primary. NRM is used in multi-drop lines connecting terminals to a host. ABM is used with a balanced configuration, and it is the most widely used transfer mode for a full-duplex point-to-point link. Either combined station may initiate a transmission without receiving the permission from the other combined station. Finally, ARM is based on an unbalanced configuration, and it is rarely used.

HDLC is a bit-oriented protocol, and it uses the frame structure shown in Figure 2.8. A single format is used for all data and control exchanges. The frame is delimited by a flag which contains the unique pattern 01111110. If frames are transmitted back-to-back, a single flag may be used to indicate the end of one frame and the beginning of the next one. Obviously, the pattern 01111110 can be easily encountered within a frame, in which case it will be interpreted as the end of the frame. To avoid this from happening, a technique known as *bit stuffing* is used. The sender always inserts an extra 0 after the occurrence of five consecutive 1's. The receiver monitors the bit stream looking for five consecutive 1's. When this pattern appears, the receiver examines the sixth bit. If it is a 0, it is deleted from the bit stream. If it is a 1 and the seventh bit is a 0, the receiver interprets the bit pattern as a delimiting flag. If the sixth bit is a 1 and the seventh bit is also a 1, then it is an error.

The second field in the HDLC frame is the address field. This is an 8-bit field used in multi-drop lines, and it is used to identify the secondary station to which the frame is transmitted. It is not necessary in a point-to-point link.

The third field in the HDLC frame is the control field. It is an 8-bit field, extendible to a 16-bit field, and its structure is shown in Figure 2.9. It is used to identify the following three types of frame: *information frame* (I-frame), *supervisory frame* (S-frame), and *unnumbered frame* (U-frame). An I-frame is used to carry data and ARQ control information, an S-frame is used to carry only ARQ control information, and a U-frame is used to provide supplemental link control functions. If the first bit of the control field is 0, then the frame is an I-frame. Otherwise, depending on the value of the second bit, it may be an S-frame or a U-frame. The meaning of the remaining sub-fields is as follows:

Flag	Address	Control	Information	FCS	Flag
8 bits	8 bits	8 or 16 bits	variable	16 or 32 bits	8 bits

Figure 2.8 The HDLC frame.

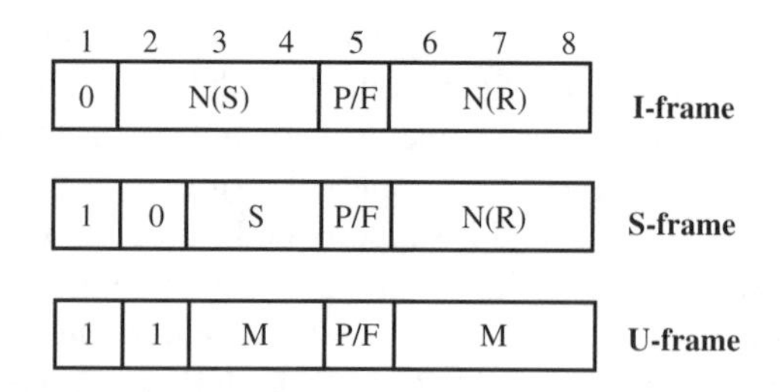

Figure 2.9 The control field of the HDLC frame.

$N(S)$: send sequence
$N(R)$: receive sequence
S: supervisory function bits
M: unnumbered function bits
P/F: poll/final bit.

During a typical exchange of information between two stations, say A and B, both stations receive and send data. This means that there are two separate ARQ mechanisms, one for the data sent from A to B and another for the data sent from B to A. The fields N(R) and N(S) in the I-frame are used to carry information for both the ARQ mechanisms piggy-backed on the frames carrying data. N(R) is used by station A to indicate to station B the current status of the ARQ from B to A, and N(S) is used by station A to indicate the sequence number of the frame that it is transmitting to B. S-frames are used when no I-frames are exchanged, and also to carry supplementary control information.

The information field is only present in the I-frames and in some U-frames. The FCS is calculated using a 16-bit CRC. A 32-bit CRC is optional.

2.5 SYNCHRONOUS TIME DIVISION MULTIPLEXING (TDM)

Time division multiplexing permits a data link to be utilized by many sender/receiver pairs, as shown in Figure 2.10. A multiplexer combines the digital signals from N incoming links into a single composite digital signal, which is transmitted to the demultiplexer over a link. The demultiplexer breaks out the composite signal into the N individual digital signals and distributes them to their corresponding output links. In the multiplexer, there

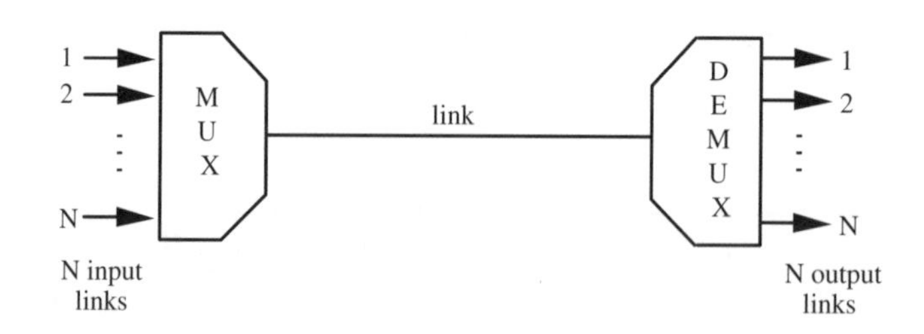

Figure 2.10 Synchronous Time Division Multiplexing (TDM).

is a small buffer for each input link that holds incoming data. The N buffers are scanned sequentially and each buffer is emptied out fast enough before new data arrives.

The transmission of the multiplexed signal between the multiplexer and the demultiplexer is organized into frames. Each frame contains a fixed number of slots, and each slot is pre-assigned to a specific input link. The duration of a slot is either a bit or a byte. If the buffer of an input link has no data, then its associated slot is transmitted empty. The data rate of the link between the multiplexer and the demultiplexer that carries the multiplexed data streams is at least equal to the sum of the data rates of the incoming links. A slot dedicated to an input link repeats continuously frame after frame, and it is called a *channel*.

TDM is used in the telephone system. The voice analog signals are digitized at the end office using the *Pulse Code Modulation* (PCM) technique. That is, the voice signal is sampled 8000 times per second, or every 125 μs, and the amplitude of the signal is approximated by a 7- or an 8-bit number. At the destination end office, the original voice signal is reconstructed from these samples. As a consequence of this sampling mechanism, most time intervals within the telephone system are multiples of 125 μs.

The standard that specifies how to multiplex several voice calls onto a single connection is known as the *digital signal level* standard, or the *DS* standard. This is a generic digital standard, and it is independent of the medium over which it is transmitted. The DS standard specifies a hierarchy of different data rates, as shown in Table 2.1. The nomenclature of this hierarchy is DS followed by the level of multiplexing. For instance, DS-1 multiplexes 24 voice channels, and it has a data rate of 1.544 Mbps. The higher levels in the hierarchy are integer multiples of the DS-1 data rate. This hierarchy is known as the *Plesiochronous Digital Hierarchy* (PDH). Plesiochronous means frame synchronous (from the Greek word *plesio*, which means frame).

The DS standard is a North American standard, and it is not the same as the international hierarchy standardized by ITU-T. Table 2.2 gives the international hierarchy, which consists of different levels of multiplexing. For instance, level-1 multiplexes 30 voice channels, and it has a data rate of 2.048 Mbps. As in the DS standard, the higher levels are integer multiples of the level-1 data rate.

The digital signal is carried over a *carrier system*, or simply a *carrier*. A carrier consists of a transmission component, an interface component, and a termination component. The T carrier system is used in North America to carry the DS signal, and the E carrier system is used to carry the international digital hierarchy. T1 carries the DS-1 signal, T2 the DS-2 signal, T3 the DS-3 signal, and so on. Similarly, E1 carries the level-1 signal, E2 carries the level-2 signal, and so on. Typically, the T and DS nomenclatures are used

Table 2.1 The North American Hierarchy.

Digital signal number	Voice channels	Data Rate (Mbps)
DS-1	24	1.544
DS-1C	48	3.152
DS-2	96	6.312
DS-3	672	44.736
DS-4	4032	274.176

Table 2.2 The international (ITU-T) hierarchy.

Level number	Voice channels	Data Rate (Mbps)
1	30	2.048
2	120	8.448
3	480	34.368
4	1920	139.264
5	7680	565.148

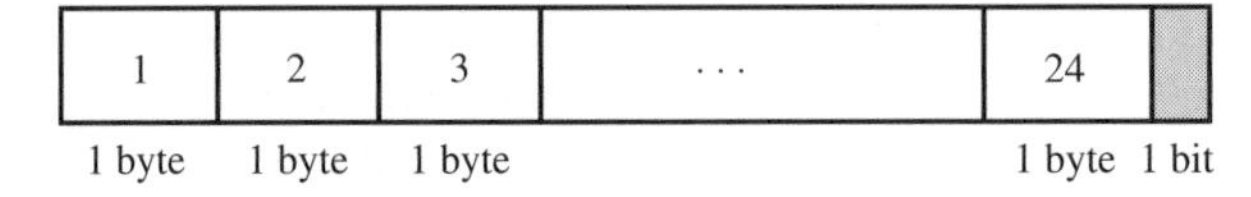

Figure 2.11 The DS-1 format.

interchangeably. For instance, one does not distinguish between a T1 line and the DS-1 signal. The same applies for the international hierarchy.

The DS-1 format, shown in Figure 2.11, consists of 24 8-bit slots and a 1-bit slot for frame synchronization. On the 1-bit slot channel, the frame synchronization pattern 1010101... is transmitted. Each of the 24 slots carries a single voice. For five successive frames, an 8-bit PCM sample is used. In the sixth frame, a 7-bit sample is used, and the 8th extra bit is used for signaling. The total transmission rate of the DS-1 format is $24 \times 8 + 1 = 193$ bits per 125 μs, corresponding to 1.544 Mbps, with each voice channel carrying a 64 Kbps voice.

The DS-1 format can be also used to carry data. In this case, 23 8-bit slots are used for data, and the remaining slot is used for control and frame synchronization. Each data slot carries 7 bits of data, amounting to a channel of 56 Kbps. The extra bit per slot is used for control.

In the international hierarchy, the level 1 format for voice consists of 32 8-bit slots, resulting in a total transmission rate of 2.048 Mbps. Of these slots, 30 are used for voice, and the remaining two are used for synchronization and control.

2.6 THE LOGICAL LINK CONTROL (LLC) LAYER

A *Local Area Network* (LAN) or a *Metropolitan Area Network* (MAN) consists of a transmission medium which is shared by all the stations that are attached to it. Access to the transmission medium is achieved through a *Medium Access Control* (MAC) protocol.

The IEEE LAN/MAN Standards Committee has produced several standards for local and metropolitan area networks, such as the IEEE 802.3 standard for Ethernet, the IEEE 802.4 standard for the token bus, and the IEEE 802.5 standard for the token ring. The *Logical Link Control* (LLC) protocol was defined in the IEEE 802.2 standard, and it runs over several different MACs.

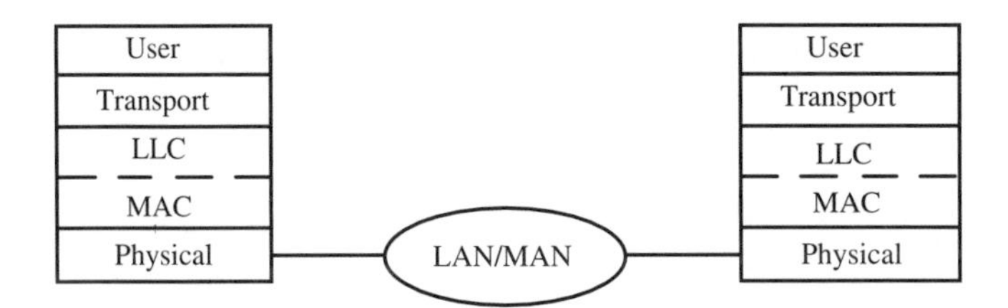

Figure 2.12 The OSI stack for LANs/MANs.

The OSI stack for stations communicating over the same LAN or a MAN is shown in Figure 2.12. The data link layer in the OSI reference model corresponds to the LLC and MAC layers. As can be seen, the networking layer is not present. Typically, layer 3 carries out functions, such as routing, addressing, flow control and error control, over a sequence of links. In a LAN or a MAN, however, there is no need for routing when transmitting between two stations which are attached to the same shared medium. The other functions of layer 3 are performed by the LLC. This considerably simplifies the OSI stack.

LLC is concerned with the transmission of link-level PDUs between two stations. Addressing in LLC is achieved by specifying the source and destination LLC users. An LLC user is typically a high-level protocol or a network management function. An LLC user address is referred to as a *Service Access Point* (SAP). The following services are provided by LLC:

- *Unacknowledged connectionless service*: this is a datagram type of service. It does not involve any flow or error control, and the delivery of data is not guaranteed.
- *Connection-mode service*: this is similar to the service offered by X.25. A logical connection is first set-up between two users before any exchange of data takes place. Flow and error control is provided.
- *Acknowledged connectionless service*: this is a service which is a cross between the above two services. Datagrams are acknowledged as in the connectionless mode service, but a logical connection is not set-up.

LLC is modeled after HDLC. It makes use of the asynchronous, balanced mode of operation of HDLC in order to support the connection-mode service. The unacknowledged connectionless service is supported using the unnumbered information PDU, and the acknowledged connectionless service is supported using two new unnumbered PDUs.

The LLC and MAC encapsulation is shown in Figure 2.13. The LLC header contains the following fields. I/G is a 1-bit field indicating whether the destination address is an individual address or a group address. DSAP and SSAP are 7-bit fields indicating the destination and source *Service Access Points* (SAP). C/R is a 1-bit field indicating whether the frame is a command or response frame. The LLC control field is identical to that of the HDLC with extended sequence numbers. The MAC header contains a MAC control field, the Destination Address (DA) and the Source Address (SA), and the MAC trailer carries the FCS value. The address of a station is the physical attachment point on the LAN.

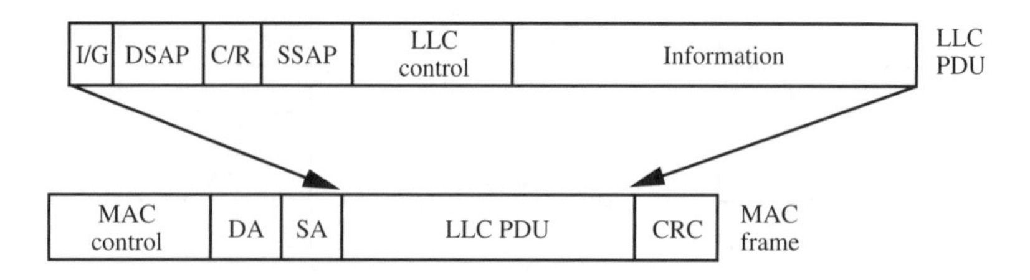

Figure 2.13 LLC and MAC encapsulation.

2.7 NETWORK ACCESS PROTOCOL X.25

X.25 was originally approved by ITU-T in 1976 to provide an interface between public packet-switched networks and their customers. Since then, it has been revised several times. It has been widely used, and it has also been employed for packet switching in ISDN. The X.25 standard specifies only the interface between a user's machine, referred to as the *Data Terminal Equipment* (DTE), and the node in the packet-switched network to which it is attached, referred to as the *Data Communication Equipment* (DCE), as shown in Figure 2.14. The standard is not concerned with the internal architecture of the packet-switched network. This is done deliberately so that vendors can use their own network architectures, while at the same time they are compatible with the end users. The standard specifies the first three layers of the ISO model. As shown in Figure 2.15, X.21 is the standard for the physical layer, LAP-B (a subset of HDLC) is the standard for the

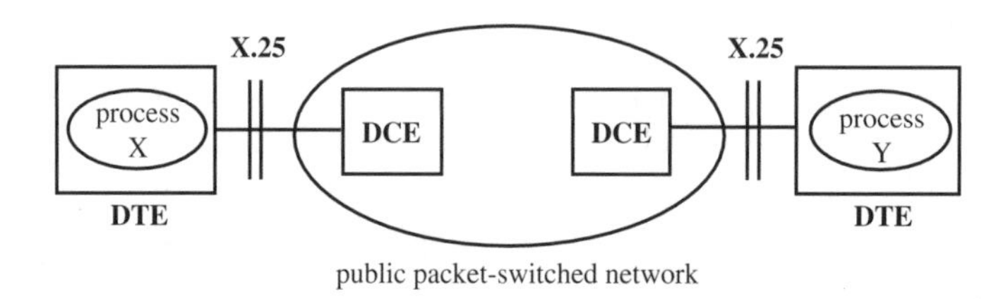

Figure 2.14 The X.25 interface.

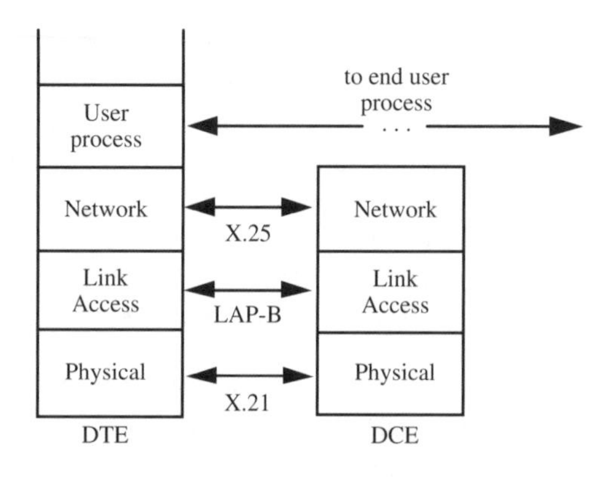

Figure 2.15 The X.25 suite.

data link layer, and X.25 is the standard for the network layer. Below, we review some of the basic features of X.25.

X.25 provides a virtual circuit service. Two types of virtual circuits are allowed: *Switched Virtual Circuits* (SVC) and *Permanent Virtual Circuits* (PVC). The following events take place in order to set-up an SVC. A pictorial view is shown in Figure 2.16.

1. The sending DTE sends a *call-request* packet to its DCE requesting to establish a virtual circuit to a specific DTE. The packet contains the source and destination addresses and a virtual circuit number selected by the DTE.
2. The network routes the call-request packet to the receiver's DCE, which sends an *incoming-call* packet to the receiving DTE. This packet has the same format as the call-request packet, but it utilizes a different virtual circuit number selected by the receiver's DTE.
3. The receiving DTE, upon receipt of the incoming-call packet, indicates acceptance of the call by sending a *call-accept* packet to its DTE using the virtual circuit number used in the incoming-call packet.
4. The network routes the packet to the sender's DCE, which sends a *call-connected* packet to the sending DTE. This packet has the same format as the call-accept packet, and it has the virtual circuit number used in the original call-request packet.
5. The sending and receiving DTEs exchange data and control packets using their respective virtual circuit numbers.
6. The sending (or the receiving) DTE sends a *clear-request* packet to terminate the virtual circuit. Its DCE sends back a *clear-confirmation* packet, and forwards the clear-request packet to the destination DCE, which issues a clear-indication packet to its DTE and from which it receives a clear-confirmation packet.

Several types of packets are used in X.25. The format for data packet with 3-bit and 7-bit sequence numbers is shown in Figure 2.17. Q is a 1-bit field which is user specified,

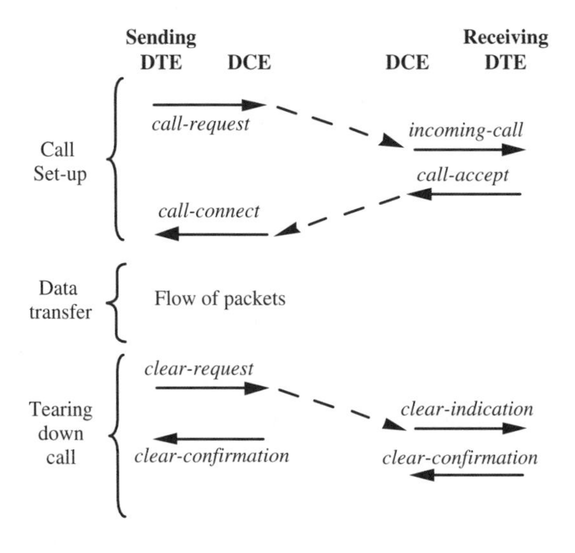

Figure 2.16 Call set-up and tearing down in X.25.

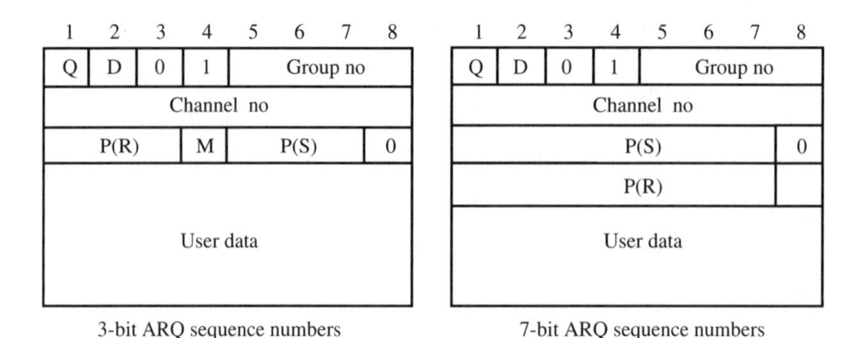

Figure 2.17 X.25 data packet formats.

and D is a 1-bit field used to indicate whether the acknowledgments are local or remote. If D = 0, the acknowledgments are between the DTE and its local DCE or the network. If D = 1, the acknowledgments come from the receiver DTE. The 12-bit field obtained by combining the fields Group no and Channel no, is used to indicate the virtual circuit number, which has local significance, i.e. it is valid only between a DTE and its local DCE. M is a 1-bit field used when packet fragmentation is employed. The P(R) and P(S) contain the receive and send ARQ sequence numbers.

2.8 THE INTERNET PROTOCOL (IP)

IP is part of the TCP/IP suite of protocols used in the Internet. TCP corresponds to the transport layer of the OSI model, and IP corresponds to the network layer of the OSI model. In this section, we describe the current version of IP, known as *IP version 4* (IPv4).

IP provides a connectionless service using packet switching with datagrams. Packets in a connectionless network, such as the IP network, are referred to as *datagrams*. An IP host can transmit datagrams to a destination IP host without having to set-up a connection to the destination, as in the case of X.25, frame relay and ATM networks. IP datagrams are routed through the IP network independently from each other, and in theory, they can follow different paths through the IP network. In practice, however, the IP network uses routing tables which remain fixed for a period of time. In view of this, all IP packets from a sender to a receiver typically follow the same path. These routing tables are refreshed periodically, taking into account congested links and hardware failures of routers and links.

IP does not guarantee delivery of IP datagrams. In view of this, if the underlying network drops an IP datagram, IP will not be aware of that. Also, IP does not check the payload of an IP datagram for errors, but it only checks its IP header. IP will drop an IP datagram, if it finds that its header is in error. Lost or erroneous data is recovered by the destination's TCP using the selective-reject ARQ scheme described in Section 2.3.

2.8.1 The IP header

An IP datagram consists of a header and a payload. The IP header is shown in Figure 2.18, and it consists of a 20-byte fixed part and an optional part which has a variable length.

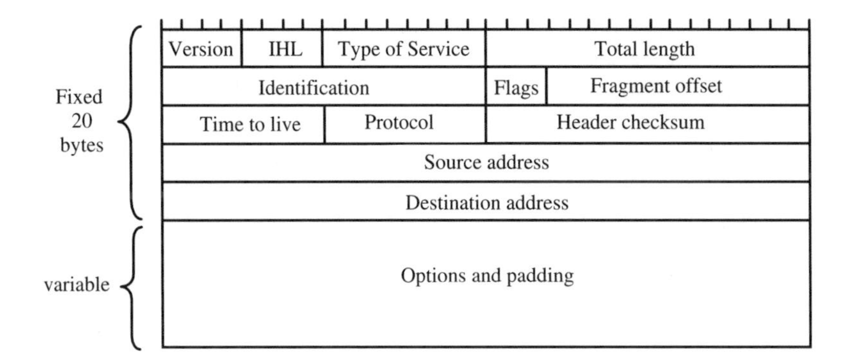

Figure 2.18 The IPv4 header.

The following fields are defined in the IP header:

- *Version*: a 4-bit field used to indicate which version of the protocol is used.
- *Internet Header Length* (IHL): this is a 4-bit field, and it gives the length of the header in 32-bit words. The minimum header length is five 32-bit words or 20 bytes.
- *Type of service*: this is an 8-bit field used to indicate whether the sender prefers the datagram to travel over a route with minimal delay or a route with maximal throughput.
- *Total length*: a 16-bit field used to indicate the length of the entire datagram, i.e. header and payload. The default value for the maximum length is 65 535 bytes.
- *Identification*: a 16-bit field used by the receiver to identify the datagram that the fragment belongs to. All fragments of a datagram have the same value in the identification field.
- *Flags*: this is a 3-bit field, but only two bits are used, namely, the 'more fragments' and the 'don't fragment'. All fragments except the last one, have the 'more fragments' bit set. This information permits the receiver to know when all the fragments have arrived. The 'don't fragment' bit is used to disallow fragmentation.
- *Fragment offset*: the 13-bit field contains an offset that points where in the original datagram this fragment belongs to.
- *Time to live*: this is an 8-bit field that specifies in seconds how long a datagram is allowed to live in the network. The maximum lifetime is 255 s. Every router that processes the datagram must decrease this field by one second, and by several seconds if the datagram is queued in the router for a long time. This field can be seen as being similar to a hop count. When the time to live field becomes equal to zero, the datagram is discarded. This prevents a datagram from moving around in the network forever.
- *Protocol*: this field is 8 bits long, and it specifies the next higher level protocol, such as TCP and UDP, to which the datagram should be delivered.
- *Header checksum*: a 16-bit field used to verify whether the IP header has been correctly received. The transmitting host adds up all the 16-bit half-words of the header using 1's compliment arithmetic, assuming that the checksum field is zero. The 1's compliment of the final result is then computed and placed in the checksum field. The receiving host calculates the checksum, and if the final result is zero, then the header has been correctly received. Otherwise, the header is erroneous and the datagram is dropped.

The checksum is recomputed at each router along the path of the datagram, since at least one field of the header (the time to live field) is changed.

- *Source address*: a 32-bit field populated with the network and host number of the sending host.
- *Destination address*: a 32-bit field populated with the network and host number of the destination host. The IP addressing scheme is discussed below.
- *Options*: a variable-length field used to encode the options requested by the user, such as security, source routing, route recording, and time stamping.
- *Padding*: a variable-length field used to make the header of the datagram an integral multiple of 32-bit words.

2.8.2 IP addresses

As we saw above, IP addresses are 32-bit long. An IP address is divided into two parts, a network and a suffix. The network identifies the physical network to which the host computer is attached, and the suffix identifies the host computer itself. The size of these two fields may vary according to the class of the IP address. Specifically, five different classes of addresses have been defined, referred to as class A, B, C, D, and E, as shown in Figure 2.19.

Classes A, B and C are called the *primary classes* because they are used for host addresses. Class D is used for multicasting, and class E is reserved for future use. The first field determines the class of the IP address, and it ranges from 1 bit for a class A address to five bits for a class E addresses. The second field gives the network address, and the third field is the suffix which gives the host address.

In class A, there is a 7-bit network address and a 24-bit host address, resulting in 128 network addresses and 16 777 216 host addresses. In class B, there is a 14-bit network address and a 16-bit host address, resulting in 16 384 network addresses and 65 536 host addresses. In class C, there is a 21-bit network address and a 8-bit host address, resulting to 2 097 152 network addresses and 256 host addresses.

Network addresses are usually written in the *dotted decimal notation*. That is, each byte is written in decimal, ranging from 0 to 255. As an example, the IP address 00000111 00000010 00000000 00000010 will be written as 7.2.0.2. Using this notation, we have that the range of class A addresses is from 1.0.0.0 to 127.255.255.255, for class

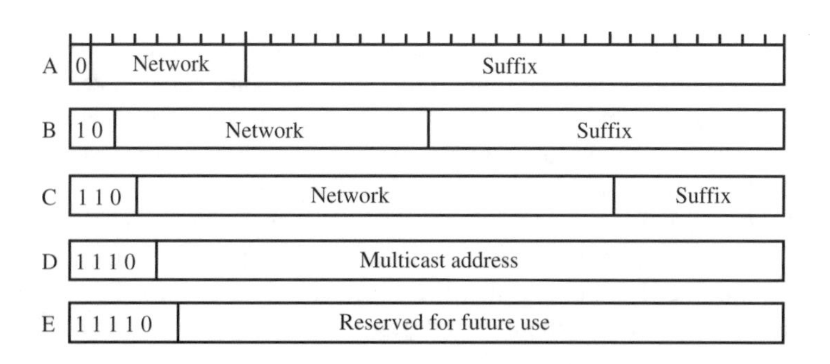

Figure 2.19 The IP address classes.

B we have a range of values from 128.0.0.0 to 191.255.255.255, and for class C we have a range of 192.0.0.0 to 233.255.255.255.

Class C is very common, whereas class A is rarely used since there are only few networks with that large number of hosts. IP reserves the host address zero to denote the address of a network. For instance, in the class B address 128.32.0.0, the network field 128.32 and the suffix is 0.0. This indicates the address of the network 128.32. For broadcasting within the network, IP uses the address 128.32.255.255.

IP assigns multiple IP addresses to routers, since a router is attached to multiple networks. Specifically, a router has one IP address for each network that it is attached to. An individual host connected to multiple networks also has multiple IP addresses, one for each network connection. Such a host is referred to as *multihomed*.

Subnetting

The IP address structure described above introduces a two-level hierarchy. The first level is the network address, and the second level is the host address carried in the suffix. In many cases, two levels of addressing is not enough. For instance, if we consider an organization with a B class address, then all the hosts appear to be organized into a single group, described by the network address. However, hosts within an organization are typically grouped together to form a number of different LANs. To distinguish the LANs, the suffix of the IP address is subdivided into a subnet part and a host part. Each LAN is assigned a subnet address carried in the subnet part, and a host in the LAN is assigned an address which is carried in the host part. The actual parsing of the suffix in these two sub-fields is dictated by a subnet mask. The subnet mask is only known to the routers within the network, since the subnets are not visible outside the network. This technique is known as *subnetting*.

Classless Inter-Domain Routing (CIDR)

In the early 1990s, it became apparent that the rapid expansion of the Internet would cause a depletion of IP addresses and an explosion of the routing tables. The main cause for the address depletion was the wasteful usage of class B addresses. Typically, an organization may have a class B address, but it may only have a small number of hosts, thus leaving the host address space largely unused. Also, the routing table explosion was due to the fact that a router is obliged to keep all the addresses of all the registered networks.

To alleviate these two problems the *Classless Inter-Domain Routing* (CIDR) scheme was proposed. This scheme permits the assignment of contiguous class C addresses, and at the same time, it reduces the number of entries required in a routing table.

The basic idea in CIDR is to allocate blocks of class C network addresses to each ISP. Organizations using the ISP are sub-allocated a block of 2^n contiguous addresses. For instance, if an organization requires 2000 addresses, then it will be allocated a block of 2048 or 2^8 contiguous class C addresses.

Hierarchical sub-allocation of addresses in this manner implies that clients with addresses allocated out of a given ISP will be routed via the ISP's network. This permits all these addresses to be advertised outside the ISP's network in an aggregate manner. As an example, let us assume that an ISP was allocated 131 072 class C network addresses starting at 194.0.0.0. That means that the lowest network address is

194.0.0.0, or 11000010 00000000 00000000 00000000, and the highest network address is 195.255.255.255, or 11000011 11111111 11111111 11111111. We observe that any address whose first seven bits are 1100001 belongs to the group of addresses allocated to the ISP. This *prefix* can be calculated by performing a bit-wise AND operation between the lowest address and the mask 254.0.0.0 or 11111110 00000000 00000000 00000000. Routers outside the ISP's network are provided, therefore, only with the base address 194.0.0.0 and the mask 254.0.0.0. This information suffices to identify whether an address of an IP packet has the same prefix as the ISP. The operation of calculating a prefix using a base network address, and a mask can be seen as the opposite of subnetting, and it is known as *supernetting*.

The above use of contiguous addresses gives rise to better usage of the address space. Also, by only advertising a base address and a mask, we minimize the amount of information that a router has to keep in its routing table. We note that there are network addresses that were allocated prior to the introduction of CIDR, and a router has to keep these addresses in its table as well.

To further simplify routing, blocks of addresses were also allocated according to geographic regions, as shown in Table 2.3.

Finally, we note that the class A, B, and C addresses are no longer used for routing. Instead, CIDR is applied to all addresses, which explains why this scheme is called classless.

2.8.3 ARP, RARP and ICMP

The TCP/IP protocol suite includes other protocols such as the *Address Resolution Protocol* (ARP), the *Reverse Address Resolution Protocol* (RARP) and the *Internet Control Message Protocol* (ICMP).

ARP is used to translate a host's IP address to its corresponding hardware address. This address translation is known as *address resolution*. The ARP standard defines two basic messages, namely a *request* and a *response*. A request message contains an IP address and requests the corresponding hardware address. A reply message contains the IP address sent in the request and the hardware address.

RARP does the opposite to ARP. It identifies the IP address of a host that corresponds to a known hardware address.

ICMP defines several error and information messages used in the Internet to report various types of errors, or send various types of information. Some of the principal messages are: *source quench*, *time exceeded*, *destination unreachable*, *redirect*, *fragmentation required*, *parameter problem*, *echo request/reply* and *timestamp request/reply*.

Table 2.3 Allocation of addresses per region.

Region	Lower address	Higher address
Europe	194.0.0.0	195.255.255.255
North America	198.0.0.0	199.255.255.255
Central/South America	200.0.0.0	201.255.255.255
Pacific Rim	202.0.0.0	203.255.255.255

A *source quench* message is sent by a router when it has run out of buffer space and it cannot accept more datagrams. A *time exceeded* message is sent by a router when the time to live field in a datagram is zero. The datagram is dropped by the router. The same message is also used by a host if the reassembly timer expires before all fragments from a given datagram have arrived. A *destination unreachable* message is sent by a router to a host that created a datagram, when it decides that the datagram cannot be delivered to its final destination. A *redirect message* is sent by a router to the host that originated a datagram, if the router believes that the datagram should have been sent to another router. A *fragmentation required* message is sent by a router to the host of a datagram, if it finds that the datagram is larger than the *Maximum Transfer Unit* (MTU) of the network over which it must be sent. The datagram is rejected by the router. A *parameter problem* message is used to indicate that an illegal value has been discovered in the IP header of a datagram. *Echo reply* and *echo request* is used to test if a user destination is reachable and alive. *Timestamp request* and *timestamp reply* are similar to the echo request/reply messages, except that the arrival time of the request message and the departure time of the reply message are also recorded.

2.8.4 IP version 6 (Ipv6)

Due to the rapid growth of the Internet, it was felt that the address space of the current IP would soon be inadequate to cope with the demand for new IP addresses. This consideration, coupled with the need to provide new mechanisms for delivering real-time traffic, such as audio and video, led to the development of a new IP, known as IPv6.

IPv6 retains many of the basic concepts from IPv4. The new features are 128-bit addresses, new header format, extension headers, support for audio and video, and extensible protocol.

PROBLEMS

1. Explain why for bursty traffic, packet switching is preferred to circuit switching.
2. Consider the use of 1000-bit frames on a 1 Mbps satellite channel with a 270 ms delay (one-way propagation delay). What is the maximum link utilization for
   (a) Stop-and-wait flow control?
   (b) Flow control with a sliding-window of 7?
   (c) Flow control with a sliding-window of 127?
3. Let us see if you remember polynomial or binary division! Calculate the FCS for the following two cases:
   (a) P = 110011 and M = 11100011
   (b) P = 110011 and M = 1110111100
4. Consider the following bit stream that has been bit-stuffed: 0111101111100111110100. What is the output stream? (The bit stream does not contain delimiters.)
5. Stations A communicates with station B using the HDLC protocol over a 56 Kbps link. A transmits frames to B continuously back-to-back, and the transmission is error-free. The payload of each frame is exactly 1500 bytes. What percent of the bandwidth of the link is used for overheads (i.e. the HDLC header and trailer).
6. In the DS-1 format, what is the control signal data rate for each voice channel?
7. In X.25, why is the virtual-circuit number used by one DTE of two communicating DTEs different from the virtual-number used by the other DTE? After all, it is the same full-duplex virtual circuit!

8. Make-up an arbitrary IP header and calculate its checksum. Now, introduce errors in the bit stream (i.e. flip single bits) so that the checksum when calculated by the receiver will fail to detect that the IP header has been received in error.
9. Consider the IP address: 152.1.213.156.
   (a) What class address is it?
   (b) What is the net and host address in binary
10. A class B IP network has a subnet mask of 255.255.0.0.
    (a) What is the maximum number of subnets that can be defined?
    (b) What is the maximum number of hosts that can be defined per subnet?

3

Frame Relay

Frame relay was originally defined by ITU-T as a network service for interconnecting *Narrowband ISDN* (N-ISDN) users. However, it very quickly became a stand-alone protocol when manufacturers of communications equipment realized its potential for transmitting data over wide area networks. It is very popular, and it is typically offered by public network operators, although it has also been implemented in private networks.

Frame relay has many common features with ATM networks, such as layer 2 switching, no error or flow control between two adjacent nodes, and a feedback-based congestion control scheme which is similar in spirit to the *Available Bit Rate* (ABR) scheme in ATM networks.

We first discuss the motivation behind the development of frame relay and its basic features. Then, we describe the *frame relay UNI*, and conclude this chapter with a discussion on congestion control. The topic of congestion control is examined in detail in Chapter 7.

3.1 MOTIVATION AND BASIC FEATURES

During the 1980s, a number of significant changes took place in the environment in which communications networks operate. Transmission facilities were replaced by digital circuits based on optical fiber. The user equipment evolved from the dumb terminal to a powerful workstation which was capable of running large networking software, and which could be directly attached to a LAN or a WAN. In parallel with these technological advances, there was an increasing demand for new bandwidth-thirsty applications, such as moving large images and videoconferencing. These new applications involved bursty traffic, and required higher throughputs and faster response times.

Bringing all the above trends together, it became evident that to support the rapid transfer rates imposed by many new applications, and to provide the required response times, new high-speed WANs were needed. Due to the bursty nature of this traffic, these new networks should be based on packet switching rather than circuit switching. Frame relay and ATM networks are two such networks. Frame relay was originally defined to run over T1 and E1 links, and ATM networks were originally defined to run over OC-3 links, i.e. 155.52 Mbps. Both frame relay and ATM networks are connection-oriented packet switching networks. Frame relay was derived by removing and rearranging some of the functionality provided by some of the OSI protocol layers. ATM networks has many similarities to frame relay, but as will be seen in the next chapter, they have a radically different architecture to frame relay and IP networks.

Older communication systems were designed based on the concept that the links between nodes were inherently unreliable. The ARQ mechanism in the data link layer was devised so that to guarantee an error-free transfer of packets over an unreliable link. The advent of fiber optics, however, made the data link layer ARQ mechanism redundant, since fiber optics links introduce very few errors. Of course, there is no guarantee that the transmissions over a fiber optics link will be always error-free. Since it was not anticipated that there will be many retransmissions, it was felt that a considerable speed-up can be gained by removing the hop-by-hop ARQ scheme and simply rely on the end devices, which had become quite intelligent, to recover packets that were either erroneously received or lost. Specifically, the recovery of these packets was left to the higher protocol layers, such as TCP, which run at the end devices. This feature was implemented in both frame relay and ATM networks.

Another interesting development had to do with the speed of computers in relation to the speeds of the communication links in a WAN. In earlier computer networks, the communications links were very slow. As a result, a node, which was basically a computer system, could run the necessary networking protocols fast enough so that it did not delay the transmission of packets on the network links. In view of this, the network links were the bottleneck in the network. With the advent of fiber-based fast transmission links, the bottleneck shifted from the communication links to computers. That is, the software that executed the necessary protocol layers could not run fast enough to keep up with the new transmission speeds. In frame relay, significant software speedups were achieved by moving some of the functionality provided by layer 3 to layer 2. Specifically, the routing decision as to which node a packet should be forwarded to, which in X.25 and in IP networks is typically carried out in layer 3, was moved down to layer 2.

In Figure 3.1 we show the path through the first three protocol layers that a packet follows as it gets switched through the X.25 network. This is contrasted in Figure 3.2, where we show how switching is accomplished in frame relay. As can be seen in Figure 3.1, the switching decision is made at the network layer and the error control and recovery is done at the link layer. The two layers introduce their own encapsulation overheads, and passing a packet from one layer to another requires moving it from one buffer of the computer's memory to another, a time-consuming process.

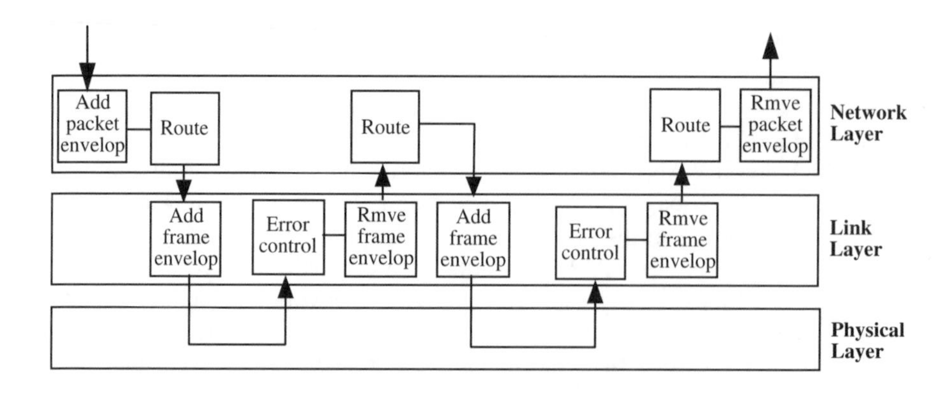

Figure 3.1 Typical flow in X.25.

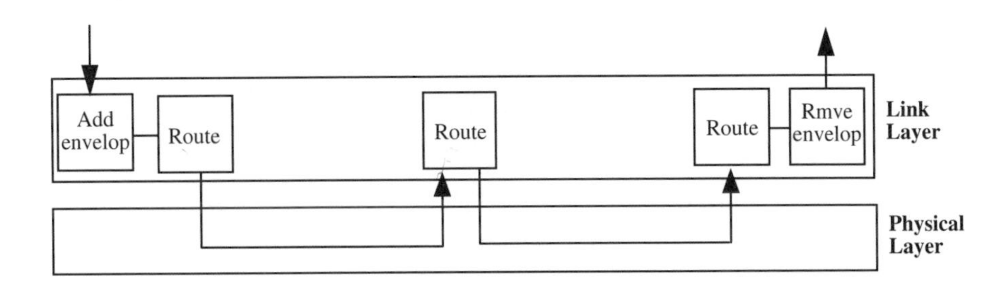

Figure 3.2 Frame relay.

In frame relay, as shown in Figure 3.2, there is no ARQ at each hop and there is no processing at layer 3. Switching is carried out within the data link layer, thus avoiding additional overheads in memory transfers and encapsulation. A frame relay packet, known as a *frame*, is discarded by a frame relay node if its header is found in error. Also, a frame can be lost if it arrives at the input buffer of a frame relay node at a time when the buffer is full. We recall that in a network equipped with hop-by-hop ARQ, erroneously received packets or lost packets in a given hop, are recovered by the local data link layer. In frame relay, discarded frames are only recovered by the end-user. As a result of these changes, frame relay provides a very efficient transport mechanisms.

As will be seen in the next chapter, ATM networks have been built on similar principles.

3.2 THE FRAME RELAY UNI

Frame relay is an extension of the *Link Access Procedure for the D channel* (LAP-D), and it is based on the core functions of ITU-T's recommendation Q.922 (Q.992 core). (LAP-D has a similar format to LAP-B, which is a subset of HDLC). LAP-D was chosen because virtually every data protocol, such as SNA, X.25, TCP/IP and DECnet, can be easily made to conform to it. Thus, frame relay is compatible with a variety of higher layer protocols.

Frame relay is a standardized *User-Network Interface* (UNI) protocol which defines the interaction between *Frame Relay Access Devices* (FRAD), such as routers, bridges, and other control devices, and a nodal processor which is known as the *frame relay network device*. Frame relay does not define how the nodal processors are interconnected. An example of the location of frame relay UNIs is given in Figure 3.3.

The protocol stack is shown in Figure 3.4, and the structure of the frame is shown in Figure 3.5. The following fields have been defined:

- *Flag*: as in HDLC, a flag consisting of the bit pattern 01111110 is used as a delimiter. Bit stuffing is used to avoid duplication of the bit pattern inside the frame.
- *Data Link Connection Identifier* (DLCI): frame relay is a connection-oriented protocol. That is, a virtual connection has to be established between the sender and the receiver before data can be transferred. The 10-bit DLCI field is used to identify a virtual connection, as explained below.
- *Command/Response* (C/R): 1-bit field used to carry application-specific information. It is carried transparently by the network.

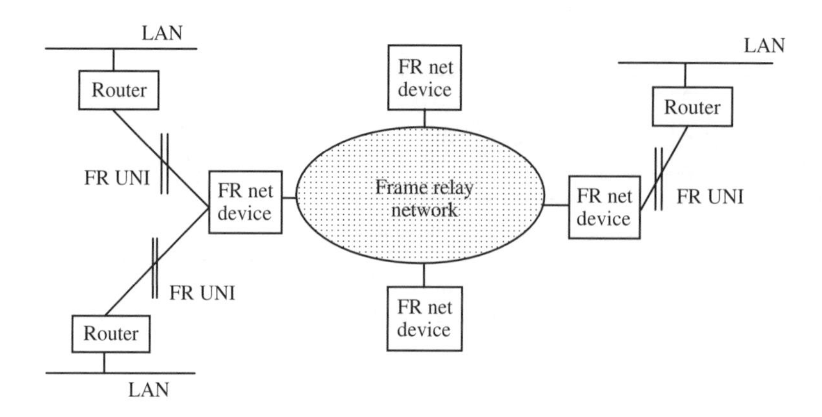

Figure 3.3 An example of the location of the frame relay UNIs.

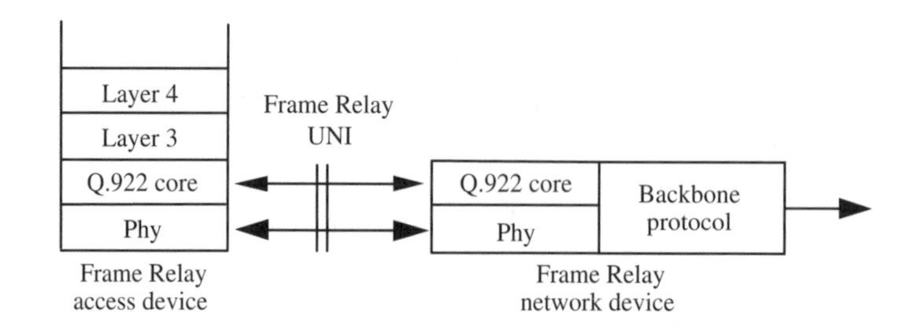

Figure 3.4 The protocol stack for the frame relay UNI.

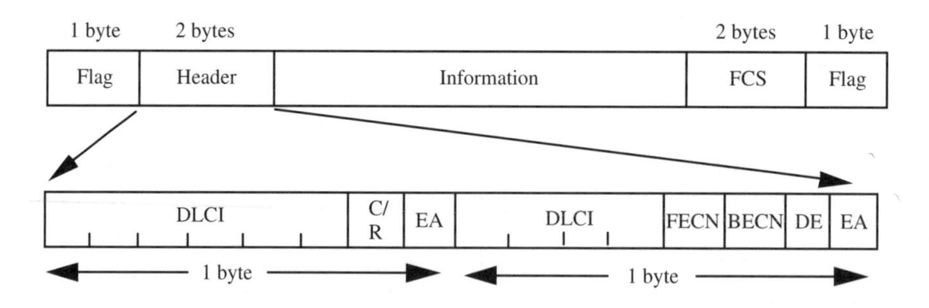

Figure 3.5 The structure of the frame relay frame.

- *Forward Explicit Congestion Notification* (FECN): 1-bit field used for congestion control, as explained in Section 3.5.
- *Backward Explicit Congestion Notification* (BECN): 1-bit field used for congestion control, as explained in Section 3.5.
- *Discard Eligibility* (DE): 1-bit field used for congestion control. The use of this bit is explained in Section 3.5.

- *Address Extension* (EA): the basic header of frame relay is two bytes. It can be extended to three or four bytes so that to support DLCIs with more than 10 bits. The 1-bit EA field is used to indicate whether the current byte in the header is the last one. For instance, in a 2 byte header, EA will be set to 0 in the first byte, and to 1 in the second byte.
- *FCS*: this field contains the frame check sequence obtained using the pattern $x^{16} + x^{12} + x^5 + 1$.

The length of the information field has to be an integer number of bytes before bit-stuffing, with a minimum size of 1 byte. The frame relay forum recommends a defaulted maximum value of 1600 bytes, and ANSI recommends up to 4096 bytes.

Frame relay is an extremely simple protocol. The receiver checks the integrity of each frame using the FCS. The frame is discarded if it is found to be in error. If no errors are found, the frame's DLCI is looked up in a table. The frame is discarded if its DLCI number is not found in the table. Otherwise, the frame is relayed towards its destination. Valid frames can also be discarded to alleviate congestion. When a frame is discarded, the end devices are not notified. A discarded frame will eventually be noted by the destination end device, which will then request the sending end device to retransmit the frame. This re-transmission procedure is not part of the frame relay protocol, and it should be done by a higher-level protocol.

DLCI: Data Link Connection Identifier

All DLCIs have local significance, i.e. they are only used to identify a virtual circuit connection between a frame relay access device and the nodal frame relay network device. The same number can be used for another connection at another UNI. This concept of using locally significant connection identifiers is also found in X.25 and in ATM networks.

In the example given in Figure 3.6, there are two connections between end devices A and B, and end devices A and C, shown by a solid line. The connection between A and B is identified by DLCI = 25 on A's UNI and by DLCI = 20 on B's UNI. The connection between A and C is identified by DLCI = 122 on A's UNI and by DLCI = 134 on C's

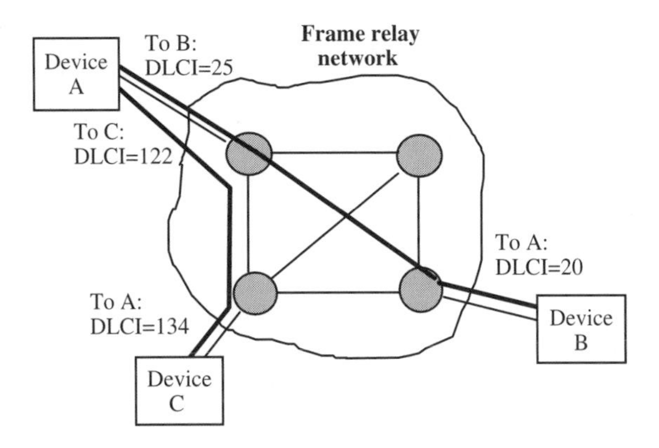

Figure 3.6 An example of DLCI addressing.

UNI. If A wants to send a frame to B, then it will populate the frame's DLCI field with the number 25. The frame relay network will deliver the frame to B with the DLCI field set to 20. Likewise, if B wants to transmit a frame to A, it will populates the frame's DLCI field with the number 20, which will be delivered to A with a DLCI equal to 25.

Frame relay connections can be Permanent Virtual Connections (PVC) or Switched Virtual Connections (SVC). PVCs are set-up administratively using network management procedures, whereas SVCs are set up on demand using the ITU-T Q.933 signaling protocol.

3.3 CONGESTION CONTROL

We recall that in a frame relay network there is no ARQ mechanism in the data link layer. Because of this, a frame will get lost if it arrives at a node during the time when the node's input buffer is full. The problem, therefore, is how to ensure a minimal frame loss while at the same time utilizing the network as much as possible. The same problem arises in an ATM network (see Chapter 7).

The data rate, expressed in bits per second, of the user's physical access channel to the frame relay network is known as the *access rate*. This is a physical upper bound on the rate at which the user sends traffic to the network, and it may take values such as 64 Kbps, 128 Kbps and 1.544 Mbps.

The traffic that a user sends to a frame relay network is described by the following three parameters: *committed burst size* (B_c), *excess burst size* (B_e) and *Committed Information Rate* (CIR). The committed burst size B_c defines the maximum number of bits a user may submit to the network during a predetermined time-interval T_c. The number of bits submitted by the user in a time-interval T_c can be exceeded by up to a certain value defined by the excess burst size B_c. Finally, the CIR is the data transmission rate, expressed in bits per second, that the frame relay network is committed to transfer, and it is calculated as the average transmission rate over the time interval T_c. These three traffic parameters are negotiated between the user and the frame relay network at set-up time. The interval T_c is calculated from B_c and CIR as follows: $T_c = B_c/\text{CIR}$.

The network polices continuously the amount of data submitted by the user during each interval T_c. At the beginning of each time interval, the cumulative number of received bits is set to 0. When a frame arrives, the cumulative number of bits received is updated, and if it is less than or equal to B_c, the frame is allowed to enter the network and the network guarantees its delivery. However, if it exceeds B_e, the network will attempt to carry this frame without any guarantee of delivering it. Finally, if it exceeds $B_c + B_e$, then the frame is discarded. This policing is carried out on each successive time-interval, which are back-to-back and non-overlapping.

An important factor in this policing scheme is the DE (Discard Eligibility) bit in the frame. The DE bit is used to indicate whether a frame can be discarded in case when congestion arises in a node. This bit can be set by the user or by the network through the policing mechanism. In the latter case, it is used by the network to mark the excessive traffic, that is the traffic that exceeds the B_c value but it is still less or equal to $B_c + B_e$. The network marks the violating frames, by setting their DE to 1. If the network experiences congestion, it attempts to alleviate it by discarding all the frames whose DE = 1. The DE bit is similar to the CLP bit used in ATM networks (see Section 4.2).

An example of this policing mechanism is given in Figure 3.7. Time is indicated on the horizontal axis, and the cumulative number of bits submitted to the network by the user in

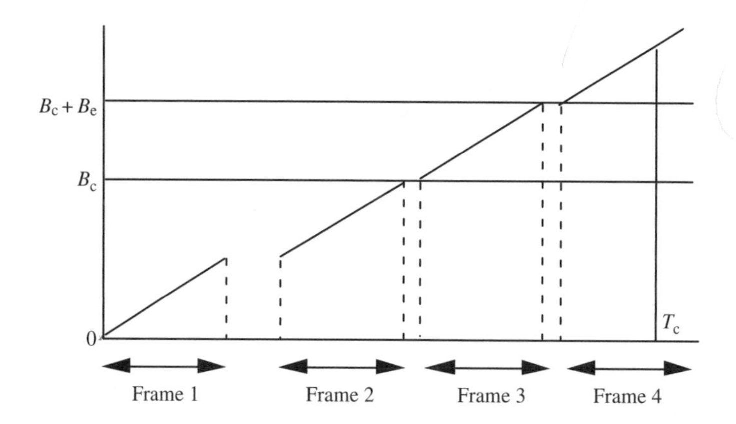

Figure 3.7 Policing the amount of traffic submitted by the user.

a single time-interval is indicated on the vertical axis. We note that on the horizontal axis, a single time-interval T_c is marked, and on the vertical axis there are two marks, one for B_c and another for $B_c + B_e$. The rate of transmission from the user to the network is equal to 1 bit per unit-time. The time it takes to transmit each frame and the cumulative number of received bits is indicated on the horizontal and vertical axes, respectively. As can be seen, the cumulative number of bits when frame 1 is received is less than B_c, and consequently the frame is allowed to enter the network and its delivery is guaranteed. Likewise, frame 2 will also enter the network and its delivery is guaranteed. Frame 3 also arrives within the same time-interval, but now the cumulative number of bits received is over B_c and exactly equal to $B_c + B_e$. In this case, frame 3 will be marked as a violating frame, that is, its DE bit will be set to 1, and will be allowed into the network, but its delivery is not guaranteed. Frame 4 also begins to arrive during the same time-interval, but since the total number of bits is over the limit, it will be simply rejected. This policing mechanism applies continuously to each successive time-interval. The time-intervals are back-to-back and they do not overlap. Also, we note that each connection is policed separately.

In addition to the above policing mechanism, the frame relay network employees a congestion control scheme inside the network, which is feedback-based. That is, when congestion occurs, the network requests the transmitting end devices to reduce their transmission rate, or even to stop transmitting, for a period of time until the congestion goes away. This scheme is known as the *explicit congestion notification*, and it could be either a *Backward Explicit Congestion Notification* (BECN), or a *Forward Explicit Congestion Notification* (FECN), as shown in Figure 3.8.

In FECN, if congestion occurs at a node, as witnessed by the occupancy level in an output buffer, the node turns on the FECN bit of all the frames going out from this output buffer. (It is possible that the EFCN bit of these frames has already been turned on, if the frames have already experienced congestion at an upstream node. In this case, no further action is taken.) All downstream nodes along the paths followed by the frames will know that congestion has occurred in an upstream node. Finally, a receiving end device will be notified that congestion has occurred in one or more upstream nodes since the FECN field of the frames that it will receive will be on. The receiving end device can then slow down the transmitting end device, using a method such as reducing its window size, or delaying

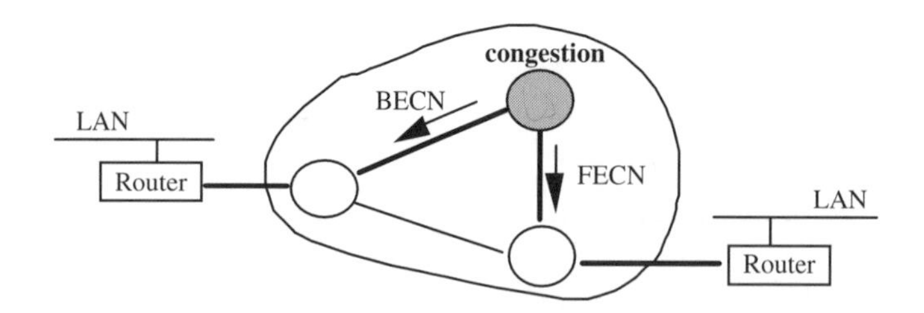

Figure 3.8 Backward and forward explicit congestion notification.

to send back acknowledgments. The action to slow down the transmitter, of course, will be taken by an upper layer protocol such as TCP. The FECN bit is the same as the EFCN bit in ATM networks (see Section 4.2).

With some upper layer protocols, it is more efficient to notify directly the transmitting end devices that congestion has occurred, rather than their receiving end devices. This can be achieved using the BECN scheme. In this scheme, the node turns on the BECN bit of the frames going to the opposite direction, i.e. towards the transmitting end devices. All upstream nodes along the path followed by these frames will know that congestion has occurred in one or more nodes downstream, and eventually, a transmitting end device will be notified.

To clarify these two schemes, we consider a generic switch shown in Figure 3.9, consisting of four input ports and four output ports, marked from 1 to 4. Each input port consists of an input link and an input buffer, where incoming frames are stored. Likewise, each output port consists of an output buffer and an output link. Incoming frames are processed by the switch, and then they are placed in the appropriate output buffer from where they get transmitted out. Processing of a frame involves the following two steps: (a) carry out the CRC check to verify that the frame's header has been correctly received; and (b) look-up the DLCI in a table to find out the new DLCI and the destination output port number from where the frame should be transmitted out. We note that Figure 3.9 gives an *unfolded* view of the switch. That is, the input ports seems to be different from the output ports. In real life, however, input port i and output port i are in fact a single duplex port.

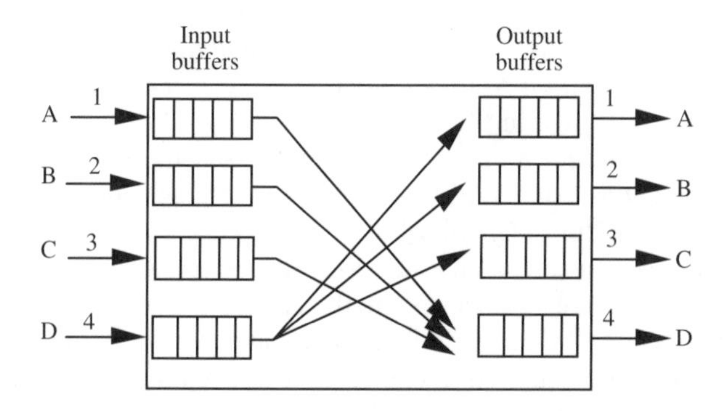

Figure 3.9 An unfolded generic switch.

Now, let us assume that end devices A, B, C and D are attached directly to this switch, as follows. A is attached to port 1, B to port 2, C to port 3, and D to port 4. That is, A transmits frames to the switch through input port 1, and receives frames from the switch from output port 1. Similarly, B transmits to the switch through input port 2, and receives frames from the switch through output port 2, and so on. We assume that A, B and C have established separate connections to D. These connection are duplex. That is, user A sends frames to the input buffer 1, which are then transferred to the output buffer 4, from where they get transmitted out to user D. User D transmits frames to A in the reverse order. That is, it sends frames to the input buffer 4, from where they get switched to the output buffer 1 and then get transmitted out to user A. The same applies to the other connections.

We are now ready to explain the two congestion control schemes. Let us assume that the output buffer 4 gets congested. That is, the total number of frames in the buffer (received from A, B and C) is over a prespecified threshold, such as 70 % of its total buffer capacity. (This threshold was selected arbitrarily for the sake of the example.) In the FECN scheme, the switch will turn the FECN bit of all the frames going out from output port 4. As a result, D will recognize that congestion has occurred upstream, and through its higher layer protocols will attempt to slow down A, B and C.

In the BECN scheme, the switch will identify which connections utilize output port 4, and turn the BECN bit of the frames going in the opposite direction to these connections. That is, it will turn on the bit of the frames going from D to A, D to B and D to C. End devices A, B and C will receive frames from D with their BECN bit turned on, they will deduce that congestion has occurred downstream from them, and they will reduce or stop their transmissions to D.

For these two schemes to work well, a number of parameters have to be tuned carefully: (a) the buffer threshold above which the switch should start turning on the FECN or BECN bit; (b) how many frames with their FECN or BECN bit turned on should be received before some action should be taken; (c) by how much a transmitting source should slow down; and (d) a buffer threshold below which the switch stops turning on the FECN or BECN bit.

The Consolidated Link Layer Management

For the BECN scheme to work, there should be frames going back towards the transmitting end devices. If there are no frames going in the reverse direction, a transmitting end device cannot be informed that congestion has occurred somewhere in the network. In view of this, ANSI and ITU-T developed the *Consolidated Link Layer Management* (CLLM), an optional signaling mechanism for congestion control. A CLLM message contains a list of DLCIs that are likely to be causing congestion. Using such messages, the network can directly request that the owner of the DLCI temporarily suspends transmission. The CLLM message follows the same length restrictions as the regular frame relay frame, and therefore it may consist of more than one frame. CLLM uses DLCI 1023. The CLLM and BECN schemes may be used together or separately.

PROBLEMS

1. Describe the main features of frame relay.
2. Explain why in frame relay there is no processing at layer 3.

3. The structure of the frame relay 3-byte header is as follows:

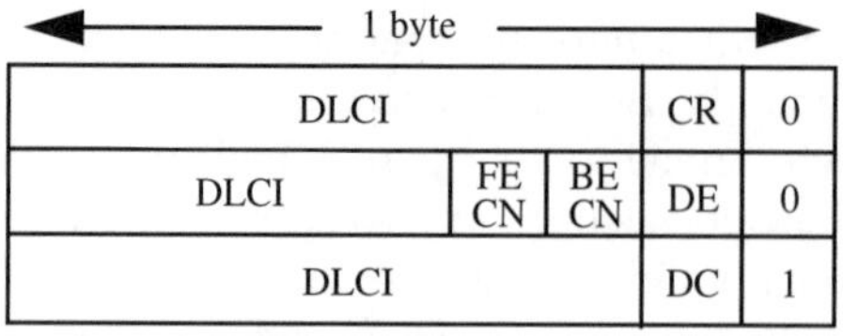

← 1 byte →

<table>
<tr><td colspan="3">DLCI</td><td>CR</td><td>0</td></tr>
<tr><td>DLCI</td><td>FE
CN</td><td>BE
CN</td><td>DE</td><td>0</td></tr>
<tr><td colspan="3">DLCI</td><td>DC</td><td>1</td></tr>
</table>

   Calculate the total number of possible DLCI values that can be defined with this header.

4. Give two reasons why it is preferable that the DLCI values have a local significance and not global.
5. A user is connected to a frame relay network over a 128 Kbps link. The user negotiates a CIR of 100 Kbps.
   (a) How does the time-interval T_c changes as B_c changes from 12 000 bits to 40 000 bits?
   (b) Explain this behavior of T_c.
   (c) How is it possible for the user to submit to the network more than B_c for each time-interval?
6. Describe in your own words how the FECN and BECN schemes work.

Part 2

The ATM Architecture

4

Main Features of ATM Networks

In this chapter, we present the main features of the ATM architecture. We start with a brief account of the considerations that led to the standardization of the ATM packet, known as a *cell*. Then, we describe the structure of the header of the ATM cell, the ATM protocol stack, and the various ATM interfaces that have been standardized. Finally, we describe the physical layer and various public and private physical layer interfaces.

4.1 INTRODUCTION

The Asynchronous Transfer Mode (ATM) is the preferred architecture for Broadband ISDN (B-ISDN). The term *Asynchronous Transfer Mode* is in contrast to *Synchronous Transfer Mode* (STM), which was proposed prior to the standardization of ATM, and which is based on the PDH hierarchy. The term *transfer mode* means a telecommunication technique for transferring information.

The ATM architecture was designed with a view to transmitting voice, video and data on the same network. These different types of traffic have different tolerance for packet loss and end-to-end delay, as shown in Table 4.1. For instance, packets containing voice have to be delivered on time so that the play-out process at the destination does not run out of data. On the other hand, the loss of some data may not necessarily deteriorate the quality of the voice delivered at the destination. At the other extreme, when transporting a data file, loss of data cannot be tolerated, since this will compromise the file's integrity, but there is no stringent requirement that the file should be delivered as fast as possible.

ATM was standardized by ITU-T in 1987. It is based on the packet-switching principle, and it is connection-oriented. That is, in order for a sender to transmit data to a receiver, a connection has to be established first. The connection is established during the call set-up phase, and when the transfer of data is completed, it is torn down. ATM, unlike IP networks, has built-in mechanisms for providing different quality-of-service to different types of traffic.

As in frame relay, there is neither error control nor flow control between two adjacent ATM nodes. Error control is not necessary, since the links in a network have a very low bit-error rate. In view of this, the payload of a packet is not protected against transmission errors. However, the header is protected in order to guard against forwarding a packet to the wrong destination! The recovery of a lost packet or a packet that is delivered to its destination with erroneous payload is left to the higher protocol layers. The lack of flow control requires congestion control schemes that permit the ATM network operator to carry as much traffic as possible without losing too many cells. These congestion control schemes are examined in Chapter 7.

Table 4.1 Different tolerances for different types of services.

	Packet loss sensitive	**Delay sensitive**
Voice	low	high
Video	moderate	high
Data	high	low

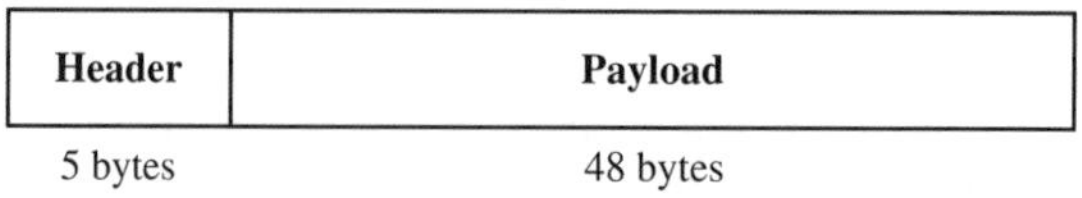

Header	Payload
5 bytes	48 bytes

Figure 4.1 The ATM cell.

The ATM packet is known as a *cell*, and it has a fixed size of 53 octets. It consists of a payload of 48 octets and a header of 5 octets, as shown in Figure 4.1. It was originally envisioned that the links in an ATM network will have a minimum speed of 155.52 Mbps, which corresponds to an OC-3 link. At such high speeds, an ATM switch does not have much time to process the information in the header. In view of this, the header provides a limited number of functions, such as virtual connection identification, a limited number of congestion control functions, payload-type indication, and header error control.

Several consideration led ITU-T to decide on a packet size of 53 octets. Below, we examine some of them.

Delay Through the ATM Network

In the early days of the standardization process of ATM, which was around the mid-1980s, it was assumed that the incoming and outgoing buffers of an ATM switch would be small, so that the queueing delay in a buffer would also be small. Telephone operators and equipment providers, who had a significant influence in the early standardization process of ATM, felt that small buffers were necessary in order to guarantee quick delivery of packets containing voice. (As will be seen in Chapter 6, this is no longer necessary, and in fact, currently ATM switches are equipped with very large buffers.) An argument, therefore, in favor of short packets had to do with the queueing delay of a packet in a buffer. If packets are short and buffers are small, then the queueing delay of a packet in a switch is also small.

Other arguments in favor of short packets were centered round the packetization delay in a real-time application. If a real-time application generates data at a rate which is very slow compared to the transmission speed of the network, then it will take a long time to fill up a packet, particularly if the packet has a large size. For instance, let us consider the transmission of 64 Kbps voice over ATM. We recall from Section 2.5, that the voice signal is sampled 8000 times per second, which gives rise to 8000 bytes per second, or one byte every 125 μs. If the packet size is 16 bytes, then it will take 16×125 μs, or 2 ms, to fill up a packet. If the packet size is 64 bytes, then it will take 64×125 μs, or 8 ms to fill up a packet. So, the smaller the packet size, the less the delay to fill up a packet. Of course, the packetization delay can be kept low if a packet is partially filled. This, however, may result in lower utilization of the network.

Echo Cancellation

To avoid echo problems for voice, the end-to-end delay has to be kept small. ITU-T requires the use of echo cancellers if the delay is greater than 24 ms. This was another argument in favor of short packets.

Header Conversion

The main function of an ATM switch is to transfer cells from its input ports to its output ports. When a cell arrives at an ATM switch on an input port, the switch has to find out its destination output port. This is done by using the connection identifier, carried in the cell's header, in a table in order to obtain the destination output port and a new connection identifier. This mechanism is known as header conversion, and it will be explained later in Section 4.2. Header conversion has to be done on the fly.

An argument in favor of longer packets was made based on the fact that the longer the packet, the more time the switch has to do the header conversion. For instance, let us assume that an ATM switch is equipped with links that transmit at OC-3, i.e. 155.52 Mbps. If the cell size is 53 bytes, then a maximum of about 365 566 cells can arrive per second. This translates to 2.7 μs per cell. That is, assuming that cells arrive back-to-back, a new cell arrives approximately every 2.7 μs. This means that the switch has 2.7 μs available to carry out the header conversion. Now, let us assume a cell size of 10 bytes. Then, a maximum of about 1 937 500 cells can arrive per second. That is, if cells arrive back-to-back, a new cell arrives approximately every 0.5 μs. In this case, the switch has only 0.5 μs for the header conversion.

Fixed vs. Variable Packet Length

The question as to whether the ATM packet should be of variable or fixed size was also debated at ITU-T. An argument in favor of variable-size packets had to do with the overhead introduced by the header. A variable-size packet was anticipated to be much larger than the small fixed-size packet that was being considered by ITU-T. Therefore, the overhead due to the header would have been much lower for variable-size packets than for fixed-size packets.

Despite this argument in favor of variable-size packets, fixed-size packets were deemed preferable because it was less complex to design ATM switches for fixed-size packets than for variable-size packets.

The Compromise!

There was an agreement in ITU-T that the ATM packet should be of a fixed size with a length between 32 and 64 bytes. The European position was in favor of 32 bytes, since there will be no need for echo cancellers. The USA/Japan position was in favor of 64 bytes, due to transmission efficiencies. In ITU-T SGXVIII meeting in 1989, the members agreed to compromise to a 48 byte payload!

4.2 STRUCTURE OF THE ATM CELL HEADER

Two different formats for the cell header were adopted, one for the UNI and a slightly different one for the *Network-Network Interface* (NNI). The UNI is concerned with the interface between an ATM end device and the ATM switch to which it is attached. An ATM end device is any device that can be attached directly to an ATM network, and it can transmit and receive ATM cells. The NNI is used between two ATM switches belonging to the same network or to two different networks.

The format of the cell header for these two interfaces is shown in Figure 4.2. As we can see, these two formats only differ in the first field. We now proceed to discuss in detail each field in the header. Understanding the meaning of these fields helps to better understand the underlying network architecture.

4.2.1 Generic Flow Control (GFC)

This field permits multiplexing of transmissions from several terminals on the same user interface. It is used to control the traffic flow from the end device to the network.

4.2.2 Virtual Path Identifier/Virtual Channel Identifier (VPI/VCI)

As mentioned in the previous section, ATM is connection oriented. An ATM connection is identified by the combined *Virtual Path Identifier* (VPI) and *Virtual Channel Identifier* (VCI). Such a connection is referred to as a *Virtual Channel Connection* (VCC). The VPI/VCI field is 24 bits in the UNI interface and 28 bits in the NNI interface. The VPI field is 8 bits in the UNI interface and 12 bits in the NNI interface. Therefore, in a UNI interface we can have a maximum of 256 virtual paths, and in a NNI interface we can have a maximum of 4096 virtual paths. In each interface we can have a maximum of 65 536 VCIs. A VPI can be assigned to any value from 0 to 255. VCI values are assigned

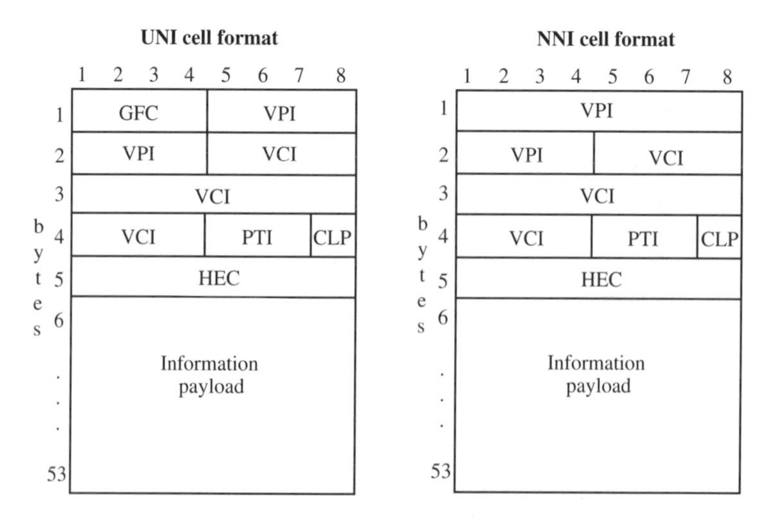

Figure 4.2 The structure of the cell header.

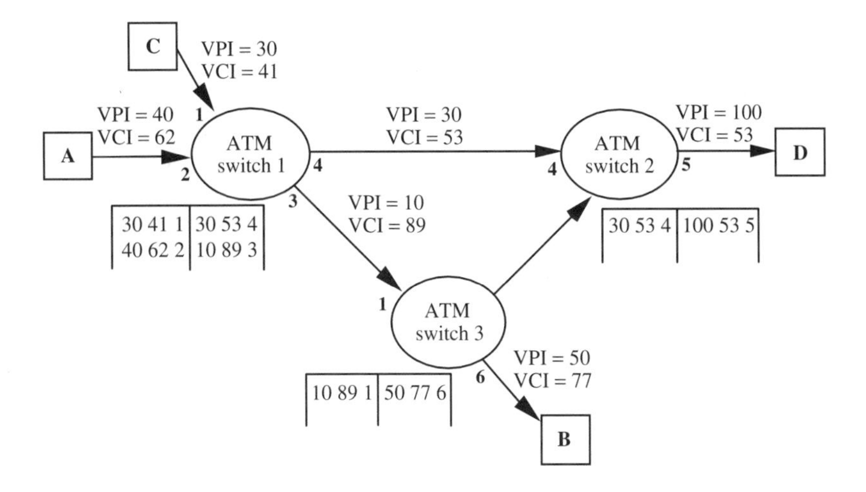

Figure 4.3 An example of label swapping.

as follows: 0–15 are reserved by ITU-T, 16–31 are reserved by the ATM Forum, and 32–65 535 are used for user VCCs.

The combined VPI and VCI allocated to a connection is known as the *Connection Identifier* (CI). That is, CI = {VPI,VCI}.

A virtual channel connection between two users consists of a path through a number of different ATM switches. For each point-to-point link that lies on this path, the connection is identified by a different VPI/VCI. That is, each VPI/VCI has *local significance*, and it is translated to a different VPI/VCI at each switch that the cell traverses. This operation is referred to as header conversion, or as *label swapping*, since the connection identifier is also known as a *label*. Label swapping (or header conversion) involves a look-up in the switching table of the ATM switch. The VPI/VCI of the incoming cell is indexed into the switching table, and the new VPI/VCI that the cell will carry in its header on the outgoing link is obtained. At the same time, the output port number is also obtained, so that the switch knows to which output port to forward the cell. As mentioned above, labels have only local significance, that is, they are only used on a single link. This simplifies the selection process of a new label for particular link.

An example of label swapping is given in Figure 4.3. Each switch is represented by a circle, and the switching table is given immediately below the circle. We assume that the switching table is centralized, and it contains information for all input ports. The first column in the switching table gives the VPI/VCI of each incoming connection and its input port. The second column gives the new label and the destination output port of each connection. Let us follow the path from A to B that traverses through ATM switches 1 and 3. We see that on the incoming link to ATM switch 1, the connection has the label VPI = 40, VCI = 62. From the switching table we find that its new label is VPI = 10, VCI = 89 and it should be switched to output port 3 of ATM switch 1. At ATM switch 3, we see that the connection's new label on the outgoing link is VPI = 50, VCI = 77, and its destination output port is 6. Therefore, the path from A to B, consists of the following three different labels: VPI/VCI = 40/62; VPI/VCI = 10/89; and VPI/VCI = 50/77. We can also identify the input and output ports of each switch, through which the connection is established. We see that it enters ATM switch 1 at input port 2, exits from the same switch from output port 3, enters ATM switch 3 at input port 1, and finally, exits from output port 6. Similarly, we can follow the path from C to D.

ATM connections are point-to-point and point-to-multipoint. Point-to-point connections are bidirectional, and point-to-multipoint connections are unidirectional. An ATM connection, depending upon how it is set-up, maybe either a *Permanent Virtual Connection* (PVC) or a *Switched Virtual Connection* (SVC). A PVC is established manually by a network administrator using network management procedures. Typically, it remains in place for a long period of time. An SVC is established in real-time by the network using signaling procedures, and it remains up for an arbitrary amount of time. The signaling protocols used to establish and release a point-to-point and point-to-multipoint SVC are described in Chapters 10 and 11.

A point-to-point SVC is established when an end device A sends a SETUP message to the switch to which it is attached, known as the *ingress* switch, requesting that a connection is established to a destination end device B. The ingress switch calculates a path through the network to B, and then it forwards the set-up request to the next-hop switch on the path, which forwards it to its next-hop switch, and so on, until the set-up request reaches the switch to which B is attached, known as the *egress* switch. This last switch sends the set-up request to B, and if it is accepted, a confirmation message is sent back to A. At that time, A may begin to transmit data. Each switch on the path allocates some bandwidth to the new connection, selects a VPI/VCI label, and updates its switching table. When the connection is terminated, each switch removes the entry in the switching table associated with the connection, the VPI/VCI label is returned to the pool of free labels, and the bandwidth that was allocated to the connection is released.

Each switch on the path of a new connection has to decide independently of the other switches whether it has enough bandwidth to provide the quality-of-service requested for this connection. This is done using a *Call Admission Control* (CAC) algorithm. CAC algorithms are examined in detail in Chapter 7.

In addition to permanent and switched virtual connections, there is another type of connection, known as a *soft PVC*. Part of this connection is permanent and part of it is switched. The connection is set-up using both network management procedures and signaling procedures.

4.2.3 Payload Type Indicator (PTI)

The field for the payload type indicator is 3 bits. It is used to indicate different types of payloads, such as user data and OAM. It also contains a bit used for explicit congestion control notification (EFCN) and a bit used in conjunction with the *ATM adaptation layer 5*. The explicit congestion control notification mechanism is similar to the FECN scheme for frame relay, discussed in Section 3.3, and it is used in the ABR scheme described in Section 7.8.1. Also, the ATM adaptation layer 5 is described in detail in Section 5.5. Table 4.2 summarizes the PTI values.

Bit 3, which is the leftmost and most significant bit, is used to indicate if the cell is a user data cell (bit set to 0), or an *Operations, Administration, Maintenance* (OAM) data cell (bit set to 1). For a user data cell, bit 2 carries the explicit forward congestion indicator. It is set to 0 if no congestion has been experienced, and to 1 if congestion has

Table 4.2 The Payload Type Indicator (PTI) Values.

PTI	Meaning
000	User data cell, congestion not experienced, SDU type = 0
001	User data cell, congestion not experienced, SDU type = 1
010	User data cell, congestion experienced, SDU type = 0
011	User data cell, congestion experienced, SDU type = 1
100	Segment OAM flow-related cell
101	End-to-end OAM flow-related cell
110	RM cell
111	Reserved

been experienced. Also, for a user data cell, bit 1 is used by the ATM adaptation layer 5. It is set to 0 if the *Service Data Unit* (SDU) type is 0, and to 1 if the SDU type is 1.

For OAM data cells, two types are defined. In addition, a *Resource Management* (RM) cell is defined which is used in conjunction with the *Available Bit Rate* (ABR) mechanism, a feedback-based congestion control mechanism described in Chapter 7.

4.2.4 Cell Loss Priority (CLP) Bit

The *Cell Loss Priority* (CLP) bit is used to indicate whether a cell can be discarded when congestion arises inside the network. If a cell's CLP bit is 1, then the cell can be discarded. On the other hand, if the cell's CLP bit is 0, then the cell cannot not be discarded.

The CLP bit is similar to the DE bit in frame relay, and its use for congestion control in ATM networks is discussed in Chapter 7.

4.2.5 Header Error Control (HEC)

The *Header Error Control* (HEC) field is used to detect single-bit or multiple-bit transmission errors in the header. CRC is used with a 9-bit pattern given by the polynomial $x^8 + x^2 + x + 1$. The HEC field contains the 8-bit FCS obtained by dividing (the first 32 bits of the header) $\times\ 2^8$ by the above pattern.

The state machine that controls the head error correction scheme is shown in Figure 4.4. It is implemented in the physical layer, as described in Section 4.5.

At initialization, the receiver's state machine is set to the *correction mode*. Each time a cell arrives, the CRC is carried out. If no errors are found, the cell is allowed to proceed to the ATM layer and the state machine remains in the correction mode. If a single-bit error is detected, then the error is corrected and the cell is allowed to proceed to the ATM layer, but the state machine switches to the *detection mode*. If a multi-bit error is detected, the cell is discarded and the state machine switches to the detection mode. In detection mode, each time a cell comes in, the CRC is carried out, and if a single-bit or a multi-bit error is detected, the cell is discarded and the state machine remains in the detection mode. If no errors are detected, then the cell is allowed to proceed to the ATM layer and the state machine shifts back to the correction mode.

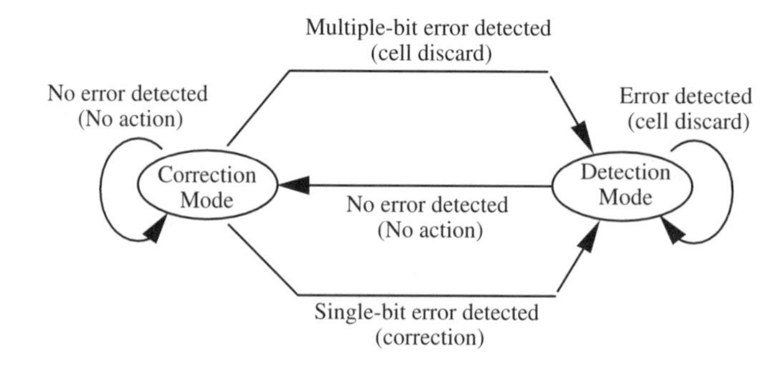

Figure 4.4 The header error control state machine.

4.3 THE ATM PROTOCOL STACK

The ATM protocol stack is shown in Figure 4.5. It consists of the physical layer, the ATM layer, the ATM adaptation layer, and higher layers that permit various applications to run on top of ATM. It is important to note that the ATM layer and the ATM adaptation layer do not correspond to any specific layers of the OSI reference model, and it is erroneous to refer to the ATM layer as the data link layer.

The ATM stack shown in Figure 4.5 is for the delivery of data. A similar stack is given in Chapter 10 for the ATM signaling protocols.

4.3.1 The Physical Layer

The physical layer transports ATM cells between two adjacent ATM layers. It is subdivided into the *Transmission Convergence* (TC) sublayer and the *Physical Medium-Dependent* (PMD) sublayer.

The TC sublayer interacts with the ATM layer and the PMD sublayer. On the transmitter's side, it receives ATM cells and generates a stream of bits that passes on to the PMD sublayer. On the receiver's side, it recovers ATM cells from the stream of bits and passes them on to the ATM sublayer. The PMD sublayer on the transmitter's side is

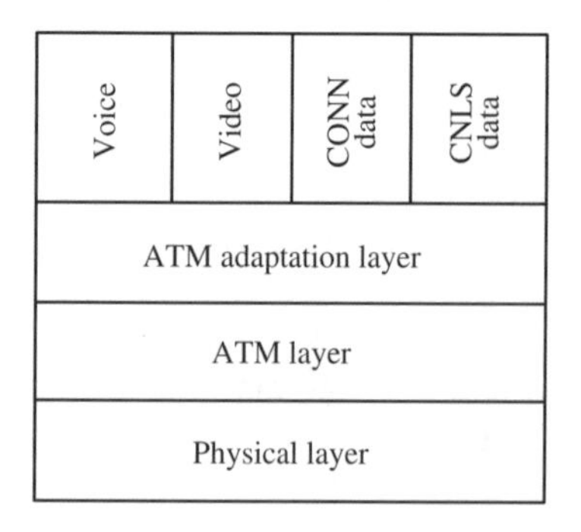

Figure 4.5 The ATM protocol stack.

concerned with the transmission and the transport across a link of a bit stream received from the TC sublayer. On the receiving side, it detects and recovers the arriving bit stream and passes it to the TC sublayer.

The physical layer and various public and private ATM physical layer interfaces are described in Section 4.5.

4.3.2 The ATM Layer

The ATM layer is concerned with the end-to-end transfer of information, i.e. from the transmitting end device to the receiving end device. Below, we summarize its main features.

Connection-Oriented Packet Switching

The ATM layer is a connection-oriented packet-switched network. Unlike the IP network, an ATM end device cannot transmit cells to a destination ATM end device over an ATM network, without first establishing a virtual channel connection. Cells are delivered to the destination in the order in which they were transmitted.

A connection is identified by a series of VPI/VCI labels, as explained in Section 4.2, and it may be point-to-point or point-to-multipoint. Point-to-point connections are bidirectional, whereas point-to-multipoint connections are unidirectional. Connections may be either Permanent Virtual Circuits (PVC) or Switched Virtual Circuits (SVC). PVCs are set-up using network management procedures, whereas SVCs are set-up on demand using ATM signaling protocol procedures.

Fixed Size Cells

In the ATM layer, packets are fixed-size cells of 53 bytes long, with a 48-byte payload and 5-byte header. The structure of the header was described in detail in Section 4.2.

Cell Switching

Switching of cells in an ATM network is done at the ATM layer. An example of the ATM stacks used when two end devices communicate with each other is given in Figure 4.6. Both end devices run the complete ATM stack, that is, the physical layer, the ATM layer,

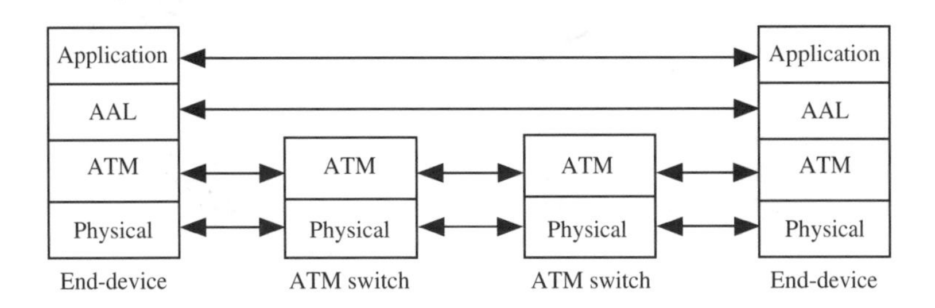

Figure 4.6 Cell switching in an ATM network.

the ATM Adaptation Layer (AAL), and the application layer. The ATM switches only need the physical layer and the ATM layer in order to switch cells. Different types of ATM switch architectures are described in Chapter 6.

No Error and Flow Control

We recall that in the OSI model the data link layer provides error and flow control on each hop using the ARQ mechanism. In ATM networks, there is neither error control nor flow control between two adjacent ATM switches which are connected with a point-to-point link. A cell simply gets lost if it arrives at an ATM switch at a time when the switch experiences congestion. Also, it is possible that the ATM network may deliver a cell to a destination end device with an erroneous payload.

The probability that a cell is delivered to the destination end device with an erroneous payload is extremely small because of the high reliability of fiber-based transmission links. Typically, the probability that a bit will be received wrongly is over 10^{-8}. Now, if we assume that bit errors occur independently of each other, then the probability that the payload of an ATM cell, which consists of 48 bytes or 384 bits, will not contain errors is $(1 - 10^{-8})^{384}$. Therefore, the probability that it will contain one or more erroneous bits is $1 - (1 - 10^{-8})^{384}$, which is very low.

The cell loss rate is a quality-of-service parameter that can be negotiated between the end device and the ATM network at set-up time. Different applications tolerate different cell loss rates. For instance, video and voice are less sensitive to cell loss than a file transfer. Cell loss rates typically vary from 10^{-3} to 10^{-6}. The ATM network guarantees the negotiated cell loss rate for each connection.

In the ATM standards there is a mechanism for recovering lost cells or cells delivered with erroneous payload, but this mechanism is only used to support the ATM signaling protocols (see SSCOP in Chapter 10). The recovery of the data carried by lost or corrupted cells is expected to be carried out by a higher-level protocol, such as TCP. We note that depending upon the application that created the data, it may not be necessary or there may not be enough time to recover such cells. For instance, it may be deemed unnecessary to recover lost cells when transmitting video over ATM.

When TCP/IP runs over ATM, the loss or corruption of the payload of a single cell results in the retransmission of an entire TCP PDU. To clarify this point, let us assume that we want to send a single TCP PDU over an ATM network. This PDU will be encapsulated by IP, and it will be passed on to the ATM network. (For simplicity, we assume no fragmentation of the IP PDU.) As will be seen in the next chapter, the ATM adaptation layer will break the IP PDU to small segments, and each segment will be placed in the payload of an ATM cell. Let us assume that the IP PDU will be carried in n ATM cells. When these n cells arrive at the destination, their payloads will be extracted and the original IP PDU will be reconstructed, from which the TCP PDU will be extracted.

Now, let us assume that one of these n cells is either lost or its payload is corrupted. If this causes the IP header to get corrupted, then IP will drop the PDU. TCP will eventually detect that the PDU is missing, and it will request its retransmission. On the other hand, if the cell in question causes the TCP PDU to get corrupted, then TCP will again detect it and it will request its retransmission. In either case, the loss of a cell or the corruption of the payload of a cell will cause the entire PDU to be retransmitted. Since this is not expected to happen very often, it should not affect the performance of the network.

Addressing

Each ATM end device and ATM switch has a unique ATM address. Private and public networks use different ATM addresses. Public networks use E.164 addresses and private networks use the OSI NSAP format. Details on ATM addresses are given in Section 10.5.

We note that ATM addresses are different to IP addresses. In view of this, when running IP over ATM, it is necessary to translate IP addresses to ATM addresses, and vice versa. Address resolution protocols are discussed in Chapter 8.

Quality of Service

Each ATM connection is associated with a quality-of-service category. Six different categories are provided by the ATM layer, namely, *Constant Bit Rate* (CBR), *Real-Time Variable Bit Rate* (RT-VBR), *Non-Real-Time Variable Bit Rate* (NRT-VBR), *Available Bit Rate* (ABR), *Unspecified Bit Rate* (UBR) and *Guaranteed Frame Rate* (GFR). The CBR category is intended for real-time applications that transmit at a constant rate, such as circuit emulation. The RT-VBR category is intended for real-time applications that transmit at a variable rate, such as encoded video and voice. The NRT-VBR category is for delay-sensitive applications that transmit at a variable rate but do not have real-time constraints. This category can be used by frame relay, when it is carried over an ATM network. The UBR category is intended for delay tolerant applications such as those running on top of TCP/IP. The ABR category is intended for applications which can vary their transmission rate according to how much slack capacity there is in the network. Finally, the GFR category is intended to support non-real-time applications that may require a minimum guaranteed rate.

Each quality-of-service category is associated with a set of traffic parameters and a set of quality-of-service parameters. The traffic parameters are used to characterize the traffic transmitted over a connection, and the quality-of-service parameters are used to specify the cell loss rate and the end-to-end delay required by a connection. The ATM network guarantees the negotiated quality-of-service for each connection.

The topic of quality of service in ATM networks is discussed in Chapter 7.

Congestion Control

In ATM networks, congestion control permits the network operator to carry as much traffic as possible without affecting the quality of service requested by the users. Congestion control may be either *preventive* or *reactive*. In preventive congestion control, one prevents the occurrence of congestion in the network using a *Call Admission Control* (CAC) algorithm to decide whether to accept a new connection, and subsequently policing the amount of data transmitted on that connection. In reactive congestion control, one controls the level of congestion in the network by regulating how much the end devices transmit through feedback messages. These two scheme are described in detail in Chapter 7.

4.3.3 The ATM Adaptation Layer

On top of the ATM layer, there is a specialized layer known as the *ATM Adaptation Layer* (AAL). The purpose of AAL is to isolate higher layers from the specific characteristics of the ATM layer. AAL consists of the *convergence* sublayer and the *segmentation-and-reassembly* sublayer. The convergence sublayer provides functions which are specific to the higher layer utilizing the ATM network. The segmentation-and-reassembly sublayer, at the transmitting side, is responsible for the segmentation of higher layer PDUs into suitable size for the information field of an ATM cell. At the receiving side, it reassembles the information fields of ATM cells into higher layer PDUs. Four different ATM adaptation layers have been standardized for the transfer of data, and they are described in Chapter 5. Also, a signaling ATM adaptation layer used by the ATM signaling protocols is described in Chapter 10.

4.3.4 Higher Level Layers

Various applications may run on top of AAL, such as voice, circuit emulation and video. Also, connection-oriented protocols (CONN) such as frame relay, connectionless protocols (CNLS) such as TCP/IP, signaling protocols and network management protocols run on top of AAL.

4.4 ATM INTERFACES

A number of interfaces have been standardized, as shown in Figure 4.7. These interfaces allow the interconnection and inter-operability of ATM equipment supplied by different vendors, such as ATM-ready workstations, ATM switches, ATM-connected routers and interworking units.

The most well-known interface is the *private user network interface* (UNI), which was one of the earliest interfaces that was standardized. This interface is commonly referred to as the UNI, rather than the private UNI, and it is concerned with the interface between an end device and a private ATM network. An ATM end device may be any device that transmits and receives ATM cells to/from an ATM network. Typically, an ATM end device

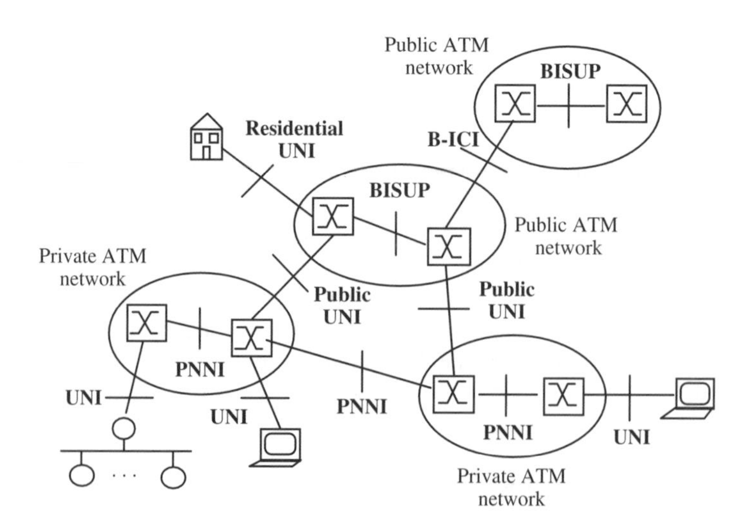

Figure 4.7 ATM interfaces.

is an ATM-ready workstation or an interworking unit that encapsulates data into ATM cells. The UNI marks the demarcation point between an ATM end device and a private ATM network. Various aspects of the UNI are described in this book. For instance, the physical layer is described in Section 4.5, the cell structure is described in Section 4.2, traffic management is described in Chapter 7, and signaling over the UNI is described in Chapter 10.

The *private network-network interface* or *Private Network Node Interface* (PNNI) is used between two ATM switches that either belong to the same private ATM network or to two different private ATM networks. This interface consists of the *PNNI routing protocol* and the *PNNI signaling protocol.* The PNNI routing protocol is responsible for routing a new call, that is a new connection, from the *calling user* to the *called user.* The calling user is the end device that issues the set-up request, and the called user is the destination end device. This connection may have to traverse several ATM switches that belong to the same network as the calling user, and it may also have to traverse several other ATM networks before it reaches the called user. The PNNI routing protocol distributes topology information between switches and networks of switches, which is used to construct paths through the network. A hierarchical mechanism enables PNNI to scale up well in large wide-area networks. The PNNI signaling protocol is used to relay signaling messages from the originating UNI to the destination UNI. PNNI is described in Chapter 11.

The public UNI is the interface used between a private ATM network or an end device and a public ATM network. This interface has been aligned with the private UNI. The interface used between two ATM switches in a public network is B-ISUP, and the interface between two public networks is the *BISDN Inter Carrier Interface* (B-ICI). A B-ICI supports an interface between carriers that allows different services, such as cell relay, circuit emulation, frame relay and SMDS.

Finally, some interesting new residential access interfaces have emerged which can be used to provide basic telephone services, known as POTS (Plain Old Telephone Service), access to the Internet, and TV channel distribution over the same wire. Typically, in the US each home has a telephone line and TV cable. The telephone line is a twisted pair that connects the home to the nearest telephone switch, commonly referred to as the *Central Office* (CO), and the TV cable is a coax cable that connects the home to the cable distribution network. The telephone network and the cable distribution network are two separate systems. One uses the telephone for POTS and also to connect to the Internet over a modem, and receives TV channels over the cable distribution network.

In recent years, cable modems have been developed that can be used to provide access to the Internet. This is in addition to TV channel distribution over the TV cable distribution network. Also, a family of technologies has been developed that can be used to provide access to the Internet, in addition to POTS, over the twisted pair. This family of technologies is known as the *x-type Digital Subscriber Line* (xDSL), where x takes different letters of the alphabet to indicate different techniques. In Chapter 9, we describe in detail one of these technologies, namely the *Asymmetric Digital Subscriber Line* (ADSL). This is a very popular technology, and it currently enables speeds from the network to the user of up to 8 Mbps, and from the user to the network of up to 800 Kbps.

Cable modems and xDSL are two competing technologies offered by different groups of operators. Cable modems are offered by cable operators and xDSL is offered by telephone operators.

4.5 THE PHYSICAL LAYER

The physical layer transports ATM cells between two adjacent ATM layers. The ATM layer is independent of the physical layer, and it operates over a wide variety of physical link types. The physical layer is subdivided into the *Transmission Convergence* (TC) sublayer and the *Physical Medium Dependent* (PMD) sublayer.

The PMD sublayer on the transmitter's side is concerned with the transmission and transport across a link of a stream of bits that it receives from the TC sublayer. At the receiver's side, it recovers the stream of bits and passes it on to the TC sublayer.

The TC sublayer interacts between the ATM layer and the PMD sublayer. On the transmitter's side, it receives ATM cells from the ATM layer and creates a bit stream that passes on to the PMD sublayer. On the receiver's side, it reconstructs the ATM cells from the bit stream that it receives from the PMD sublayer and passes them on to the ATM layer.

4.5.1 The Transmission Convergence (TC) Sublayer

The following are the main functions performed by this sublayer.

HEC Cell Generation and Verification

The ATM layer passes to the physical layer ATM cells for transmission over the link. Each ATM cell is complete, except for the HEC byte. This byte is computed and inserted into the HEC field in the TC sublayer. At the receiving side of the link, the HEC state machine is implemented in the TC sublayer. TC will drop any cell whose headers was found to be in error.

Decoupling of Cell Rate

The PMD sublayer expects to receive a continuous stream of bits. During the time that ATM cells are not passed down from the ATM layer, TC inserts idle cells in-between the cells received from the ATM layer so that to maintain the continuous bit stream expected from PMD. These idle cells are discarded at the receiver's TC sublayer. They are identified uniquely since their header is marked as: VPI = 0, VCI = 0, PTI = 0 and CLP = 0.

Cell Delineation

Cell delineation is the extraction of cells from the bit stream received from the PMD sublayer. The following procedure for cell delineation is based on the HEC field.

Let us consider a bit stream, and assume that we have guessed correctly the first bit of a new cell. This means that this bit and the following 39 bits make up the header of the cell. Consequently, if we carry out the CRC operation on these 40 bits, the resulting FCS should be zero. If it is not zero, then that means that the bit we identified as the beginning of the cell is not really the first bit. We repeat this process starting with the next bit, and so on, until we get a match. At that moment, we have correctly identified in the bit stream the beginning of a cell.

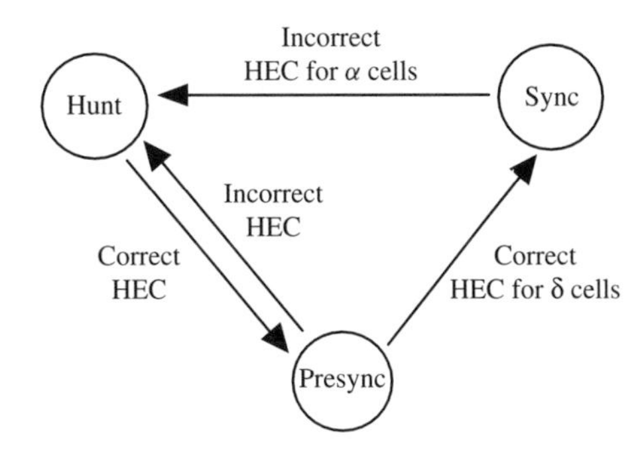

Figure 4.8 The cell delineation state machine.

This simple idea is used in the state machine for cell delineation shown in Figure 4.8. The state machine consists of the *hunt* state, the *presync* state and the *sync* state. At the beginning, the state machine is in the *hunt* state. In this state, the incoming bit stream is continuously monitored to detect the beginning of a cell using the above procedure. When a match occurs, the state machine moves to the *presync* state. In this state, it checks that the FCS of δ consecutive cells is zero. If a mismatch is found, that is the FCS of one of these cells is not zero, then the state machine goes back to the hunt state. Otherwise, synchronization with the bit stream has been achieved, and the state machine moves to the *sync* state. Synchronization is assumed to be lost if α consecutive mismatches occur. In this case, the state machine shifts to the hunt state. ITU-T recommends that $\delta = 6$ and $\alpha = 7$.

While in the sync state, the HEC state machine described in Section 4.2, is used to detect errors in the header of the incoming cells. We recall that when the state machine is in the correction mode, a cell is dropped if more than one erroneous bits are detected in its header. If only one erroneous bit is detected, then it is corrected and the cell is delivered to the ATM layer. When the state machine is in the detection mode, a cell is dropped if one or more erroneous bits are detected in its header. When the state machine is in either state, a cell is delivered to the ATM layer if no errors are detected in its header.

Transmission Frame Generation and Recovery

In frame oriented transmission systems, such as SONET, at the sender's side TC generates frames by placing frame-related information and ATM cells into a well-defined frame structure. At the receiver's side, it recovers the frames and subsequently the ATM cells from the bit stream.

4.5.2 The Physical Medium-Dependent (PMD) Sublayer

The following are the main functions performed by this sublayer.

Timing Function

This is used to synchronize the transmitting and receiving PMD sublayers. It generates timing for transmitted signals and it derives correct timing for received signals.

Encoding/Decoding

PMD may operate on a bit-by-bit basis, or with a group of bits as in the 4B/5B and 8B/10B schemes. In the 4B/5B encoding scheme, each group of 4 bits is coded by a 5-bit code, and in 8B/10B each group of 8 bits is coded by a 10-bit code. Coding groups of bits is known as *block-coding*. Block coding requires more bandwidth than it effectively provides. For instance, FDDI uses 4B/5B block coding and it runs at 125 Mbps, which gives an effective bandwidth of 100 Mbps. There are several benefits with block coding, such as bit boundary detection and exchange of control information. In addition, a speed-up in the execution of the protocol is achieved, since it operates in chunks of bits. For instance, since FDDI uses 4B/5B coding, it operates in chunks of 4 bits.

4.5.3 ATM Physical Layer Interfaces

ATM was originally developed for high-speed fiber-based links, such as OC-3. Currently, a large variety of physical layer interfaces are in use. A list of public and private interfaces is given in Tables 4.2 and 4.3.

SONET/SDH

SONET stands for *synchronous optical network*, and it was designed originally by Bellcore (now Telcordia) for the optical fiber backbone of the telephone network. An outgrowth of SONET was standardized by ITU-T, and it is known as the *Synchronous Digital Hierarchy* (SDH).

The basic structure of SONET is an 810-byte frame, which is transmitted every 125 μs. This corresponds to 51.84 Mbps, and each byte provides a 64 Kbps channel. This basic structure is referred to as the *synchronous transport signal level 1* (STS-1). Faster speeds are obtained by multiplexing several STS-1 frames, such as, STS-3, STS-12, STS-24 and STS-48. In general an STS-N signal operates at Nx51.84 Mbps.

Table 4.2 Public Interfaces.

Standard	Bit rate	Physical media
SONET STS-48	2.4 Gbps	Single-mode fiber
SONET STS-12	622.080 Mbps	Single-mode fiber
SONET STS-3c	155.520 Mbps	Single-mode fiber
SONET STS-1	51.840 Mbps	Single-mode fiber
PDH E4	139.264 Mbps	Coaxial pair
PDH DS3	44.736 Mbps	Coaxial pair
PDH E3	34.368 Mbps	Coaxial pair
PDH E2	8.448 Mbps	Coaxial pair
PDH J2	6.312 Mbps	Twisted pair/Coax
PDH E1	2.048 Mbps	Twisted pair/Coax
PDH DS1	1.544 Mbps	Twisted pair
inverse mux	nx1.544 Mbps	Twisted pair
inverse mux	nx2.048 Mbps	Twisted pair/Coax

Table 4.3 Private Interfaces.

Standard	Bit rate	Physical media
SONET STS-48	2.4 Gbps	Single-mode fiber
SONET STS-12	622.080 Mbps	Multi-mode fiber
SONET STS-12	622.080 Mbps	Single mode fiber
SONET STS-3c	155.520 Mbps	UTP 5
SONET STS-3c	155.520 Mbps	UTP 3
SONET STS-3c	155.520 Mbps	SM, MM fiber/ Coax
Fiber channel	155.520 Mbps	Multi-mode fiber
TAXI (FDDI)	100.000 Mbps	Multi-mode fiber
SONET STS1	51.840 Mbps	SM, MM fiber/Coax
SONET STS1	51.840 Mbps	UTP 3
ATM 25	25.600 Mbps	UTP 3
Cell stream	25.600 Mbps	UTP 3
Cell stream	155.52 Mbps	Multi-mode fiber/STP

SONET signals may be transported by either electrical or optical means. The STS electrical signals, when transmitted over fiber, are converted to a corresponding optical signal called the *Optical Carrier* (OC). The OC hierarchy starts at 51.84 Mbps (OC-1). Higher rates are obtained by multiplexing several OC-1s, such as OC-3, OC-12, OC-24 and OC-48.

The basic frame of SDH is known as the *synchronous transport module level 1* (STM-1), and it corresponds to 155.52 Mbps. This structure is compatible with STS-3c, which is obtained by concatenating 3 STS-1 signals. Faster speeds are obtained by multiplexing STM-1 frames into a larger frame. STM-N carries *N* times the payload of an STM-1 frame. Table 4.4 gives the SONET/SDH hierarchy.

The SONET/SDH frame structure is arranged in rows and columns, as shown in Figure 4.9. Each frame is transmitted row by row every 125 μs. The SONET/SDH overheads occupy the first few columns of the frame, and the remaining columns are used to carry the payload. Frames consist of nine rows, and in the case of STS-1, it consists of 90 columns, of which three are used for overhead. The STM-1 frame consist of 270 columns of which nine are used for overhead.

Table 4.4 The SONET/SDH hierarchy.

Optical carrier level	SDH level (ITU-T)	SONET level (electrical)	Data rates
OC-1		STS-1	51.84 Mbps
OC-3	STM-1	STS-3	155.52 Mbps
OC-12	STM-4	STS-12	622.08 Mbps
OC-24	STM-8	STS-24	1.244 Gbps
OC-48	STM-16	STS-48	2.488 Gbps
OC-N	STM-N/3	STS-N	N*51.84 Mbps

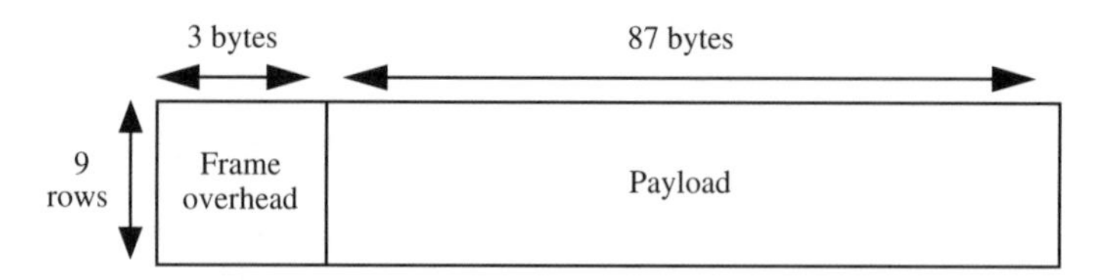

Figure 4.9 The STS-1 frame structure.

ATM cells are directly mapped onto the payload of the STM-1 frame. It is possible that an ATM cell will straddle over two frames. The start of the first cell is identified by a pointer in the overhead section. An alternative scheme for identifying the beginning of the first cell uses the cell delineation procedure described in Section 4.5.1.

Plesiochronous Digital Hierarchy (PDH)

PDH is the older digital hierarchy of signals, based on time-division multiplexing (see Section 2.5). ATM was originally defined for SONET/SDH data rates. However, it was also defined over PDH to expedite initial deployment of ATM, and also because many corporations did not require 155 Mbps or more bandwidth. ATM over PDH is deployed at both DS1/E1 and DS3/E3 data rates. Higher data rates have also been defined.

Nx64 Kbps

At data rates below DS1/E1, the ATM cell interface suffers from an unacceptable level of overheads due to the ATM cell header, and also due to the AAL encapsulation. For fractional T1 speeds (i.e. for speeds which are integer multiples of 64 Kbps), a frame-based ATM interface known as *frame UNI* (FUNI) is deployed. It is based on an older interface known as ATM-DX1, and it supports signaling, quality of service, and network management. FUNI is used for data traffic.

Inverse Mulitplexing for ATM (IMA)

Inverse multiplexing for ATM provides a middle ground for users who require multiple T1/E1 links across a wide area, but not as much as a T3/E3 link.

We recall that a multiplexer combines the traffic streams from *N* incoming links into a single data stream, which is transmitted to a demultiplexer over a link. The demultiplexer breaks out the data stream into the *N* individual data streams, and transmits them out over their corresponding output links. An example of a multiplexer/demultiplexer is shown in Figure 2.10.

IMA does the exact opposite. As shown in Figure 4.10, a demultiplexer is connected to a multiplexer with several T1/E1 links. The demultiplexer breaks up the incoming stream and transmits the ATM cells over the T1/E1 links. At the other side of these links, the multiplexer combines the individual cell streams into the original cell stream and in the order in which they arrived. In the example in Figure 4.10, we see that incoming cells 1, 2, 3, 4 and 5 are transmitted over the three links, and at the other side they are multiplexed back into a single stream with the cell order preserved.

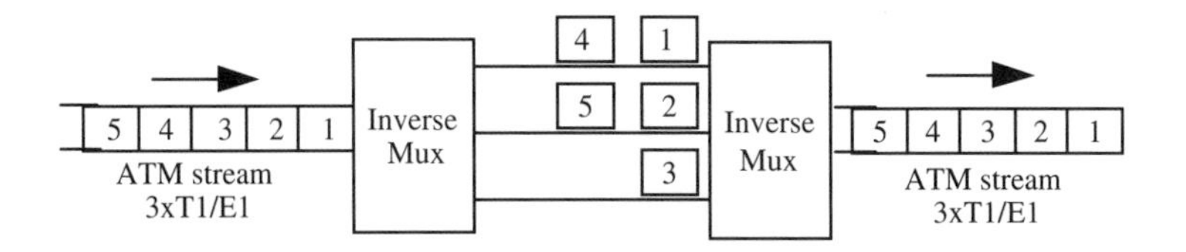

Figure 4.10 Inverse multiplexing for ATM.

The IMA protocol can also add/delete T1/E1 links on demand, compensate for different link delays, handle link failure and automatic link recovery, and support all the ATM quality-of-service categories. IMA is available for private and public interfaces, and it can also support the use of leased lines and dial ISDN lines.

TAXI (FDDI)

This interface is referred to as TAXI because of the chipset used. It was the first ATM interface to be deployed in the local area. It is equivalent to the FDDI interface, and it uses 4B/5B block coding. As mentioned in Section 4.5.2, 4B/5B block coding is based on a scheme where chunks of 4 bits are coded as 5-bit chunks. The overall data rate is 125 Mbps, of which 100 Mbps is effectively used. Unlike the single-mode fiber used by FDDI, this interface is based on multi-mode fiber, and it is limited to a distance of 2 kilometers.

ATM 25

Developed to bring affordable ATM to the desktop, it was proposed originally by IBM and Chipcom. It uses IBM's token ring physical layer, but with 4B/5B encoding.

Fiber Channel

The fiber channel standard was adopted by the ATM Forum as a method for transporting data in a local area environment. Though standardized early, it has not seen wide support. Distances range from 2 meters to 2 kilometers at data rates from 100 Mbps to 4 Gbps. Block coding 8B/10B is used.

Cell-Based Clear Channel

The clear channel interface can support multiple physical media and data rates in a local area environment. An ATM cell stream is directly transported over a number of different interfaces. Cell rate decoupling may be based on either idle or unassigned cells. Cell delineation is performed using the cell delineation state machine described in Section 4.5.1.

Other Interfaces

Interfaces for residential access networks, such as the Asymmetric Digital Subscriber Line (ADSL), wireless ATM and ATM over satellite have also been developed.

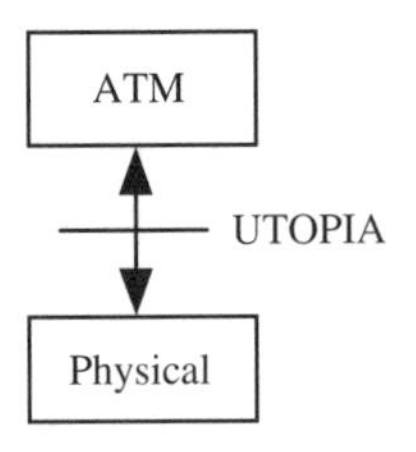

Figure 4.11 UTOPIA.

ADSL is one of the access technologies that can be used to convert the telephone line into a high-speed digital link. It is part of a family of technologies known as xDSL, where x takes different letters of the alphabet to indicate different transmission techniques. ADSL can support basic telephone service and access to the Internet. Currently, it can provide data rates of up to 8 Mbps from the network to the home, and up to 800 Kbps from the home to the network. ADSL is described in Chapter 9.

4.6 UTOPIA AND WIRE

The *Universal Test and Operations Physical Interface for ATM* (UTOPIA) is an interface between the physical layer and the ATM layer, as shown in Figure 4.11. It was defined by the ATM Forum to enable interoperability between ATM physical layer devices and ATM layer devices made by different vendors.

Various specifications for UTOPIA have been defined: UTOPIA levels 1 and 2 describe an 8-bit and 16-bit data path with a maximum data rate of 800 Mbps; UTOPIA level 3 defines an 8-bit, 16-bit and 32-bit data path with a maximum data rates of up to 3.2 Gbps; and UTOPIA level 4 defines an 8-bit, 16-bit and 32-bit data path with a maximum speed of up to 10 Gbps. Various configurations can be supported that enable the interconnectivity of one more ATM layer devices with one or more physical layer devices.

The *Workable Interface Requirements Example* (WIRE) describes an interface between the TC sublayer device and the PMD sublayer device. This is a natural interface as this is the point of change of technology from digital technology supporting moderate clock rates on the TC side to high-speed mixed mode technology on the PMD side.

PROBLEMS

1. Why is there error control for the header and not for the payload of an ATM cell?
2. Why is there no data link layer flow control in ATM networks?
3. Give one argument in favor of fixed-size ATM cells.
4. How long does it take to transmit an ATM cell over a link, when the link is:
   (a) a T1 line
   (b) an OC-3
   (c) an OC-12
   (d) an OC-24
   (e) an OC-48

5. Consider the HEC mechanism. Let p be the probability that a bit is received in error.
   (a) With what probability a cell is rejected when the HEC state machine is in the correction mode?
   (b) With what probability a cell is rejected when the HEC state machine is in the detection mode?
   (c) Assume that the HEC state machine is in the correction mode. What is the probability that n successive cells, where $n \geq 1$, will be rejected.
   (d) Assume that the HEC state machine is in the correction mode. What is the probability $p(n)$ that n successive cells will be accepted, where $n \geq 1$. (Hint: Write down the expression for $p(1)$ and $p(2)$, and express $p(3)$ as a function of $p(1)$ and $p(2)$. Then, write down the general expression for $p(n)$ for any n as a function of $p(n-1)$ and $p(n-2)$.)

6. Let us a consider a large ATM switch, and assume that the routing table is centralized and that it contains information for all input ports. To reduce the time it takes to look-up an entry in the routing table, a hardware mechanism known as *associative registers* is used. Each register contains a *key* and a *value*, where the key is the label and port number of an incoming cell, and the value is the new label and the output port of the cell. When the associative registers are presented with an item, it is compared with all the keys simultaneously. This makes the search very fast. We assume that only part of the routing table is in the associative registers, since the hardware can be expensive. The entire routing table is kept in the main memory. If an item offered to the associative registers is found, then we have a *hit*. Otherwise, we have a *miss*, and the appropriate entry from the routing table in the main memory is fetched and loaded into the associative registers by over-writing another entry.

   Let A be the number of the associative registers, T the maximum number of entries in the routing table, S the time it takes to search the associative registers, and R the average time to search the routing table and load the correct entry into the associative registers.
   (a) Write down the expression for the time W that it takes to find the outgoing label and port number, as a function of the above parameters, assuming that both the associative registers and the routing table are full.
   (b) What is the region of feasible values of A and R for which W is less than 100 ms, assuming that $T = 2000$ and $S = 0$.

7. Consider the case of an application running over an ATM network. Assume that each packet generated by the application is carried by n ATM cells which are transmitted back-to-back. The time to transmit a cell is T and the average time it takes for a cell to traverse the ATM network and reach the destination is D. When all the ATM cells belonging to the same packet are received by the destination, their payloads are extracted from the cells and they are assembled to the original packet. Subsequently, the CRC operation is carried out on the reassembled packet. If the CRC check is correct, the packet is released to the application. Otherwise, a negative acknowledgment is sent back to the source requesting the retransmission of the entire packet. The time it takes to carry out the CRC check is F, and the time it takes for the negative acknowledgment to reach the source is D. Let p be the probability that the ATM cell is received with erroneous payload.
   (a) What is the probability that all n cells are received correctly?
   (b) What is the probability that exactly m cells are received erroneously, where $m < n$.
   (c) Write down the expression for the average time, call it W, required to transfer correctly an application packet to the destination.
   (d) Plot W against the probability p, assuming that $n = 30$, $D = 20$ ms, $T = 3$ μs, and $F = 0$. Vary p from 0.1 to 0.00001.
   (e) Discuss the results of the above plot in the light of the fact that there is no data link layer in ATM.

5

The ATM Adaptation Layer

The purpose of the *ATM Adaptation Layer* (AAL) is to isolate a higher layer from the specific characteristics of the ATM layer. In this chapter, we describe the four ATM adaptation layers that have been standardized, namely, *ATM Adaptation Layer 1* (AAL 1), *ATM Adaptation Layer 2* (AAL 2), *ATM Adaptation Layer 3/4* (AAL 3/4), and *ATM Adaptation Layer 5* (AAL 5). These ATM adaptation layers are used for the transfer of user data. A fifth adaptation layer used by the ATM signaling protocols, the *Signaling ATM Adaptation Layer* (SAAL), has also been standardized, and is described in Chapter 10.

5.1 INTRODUCTION

The ATM Adaptation Layer (AAL) is sandwiched between the ATM layer and the higher-level layers, as shown in Figure 5.1. AAL consists of two sublayers, namely, the *Convergence Sublayer* (CS) and the *Segmentation And Reassembly sublayer* (SAR), as shown in Figure 5.2. The convergence sublayer provides service-specific functions, and it is further subdivided into the *Service Specific Convergence Sublayer* (SSCS) and the *Common Part Sublayer* (CPS). The segmentation and reassembly sublayer, at the transmitting side, provides segmentation of higher layer PDUs into a suitable size for the information field of an ATM cell. At the receiving side, it reassembles the information fields of ATM cells into higher layer PDUs.

The various types of services provided by AAL to higher-level applications were classified into four different classes, A, B, C and D. This classification scheme was based on the following three attributes: (a) the need to transmit timing information between source and destination; (b) whether the transmission is constant bit rate or variable bit rate; and (c) whether it is connection-oriented or connectionless. Table 5.1 summarizes the values of these three attributes for each class.

In certain cases, it is necessary for the transmitting application to be synchronized with the receiving application. To achieve such a synchronization, it may be necessary that timing information is exchanged between the two applications. The first attribute, therefore, indicates whether such an exchange of timing information is necessary. The second attribute has to do with the rate at which the source transmits, and specifically, whether it transmits at a constant bit rate or variable bit rate. A constant bit rate source produces a fixed number of bits each unit of time, whereas a variable bit rate source transmits at a rate that varies over time. For instance, a 64 Kbps voice always produces 8 bits every 125 μs, whereas the transmission rate for encoded video is variable, since the amount of information that has to be transmitted for each image varies. Another example of a variable bit rate application is the transfer of a data file. In this case, the application

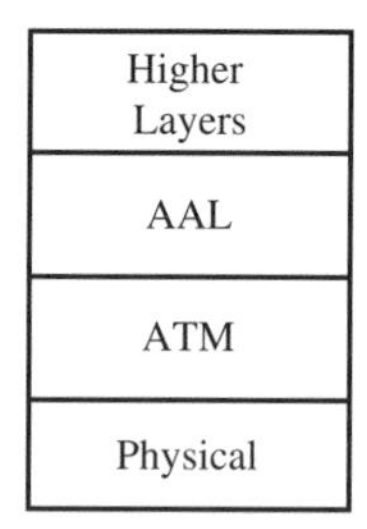

Figure 5.1 The ATM adaptation layer.

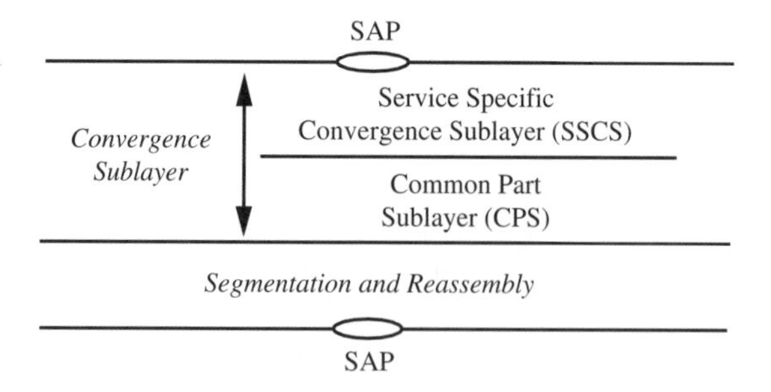

Figure 5.2 The ATM adaptation sublayers.

Table 5.1 The attributes of the four AAL classes.

Attributes	Class A	Class B	Class C	Class D
Timing between source and dest.	Yes	Yes	No	No
Bit Rate	Constant	Variable	Variable	Variable
Connection mode	CO	CO	CO	CL

does not transmit continuously. Typically, it goes through an active period of time during which it transmits data, followed by a silent period of time during which it does not transmit at all. The silent periods are due to a variety of reasons, such as the application is waiting for more data to become available, or it is waiting for an acknowledgment before it can send more data. This cycle of an active period followed by a silent period repeats until the entire file is transmitted. The application, therefore, is either transmitting or it is silent, which means that its rate of transmission varies over time. Finally, the classification between connection-oriented and connectionless mode refers to whether a source requires a connection-oriented service or a connectionless service, the implication being that the connection-oriented service has a better quality of service than the connectionless service.

A good example of class A service is the *Circuit Emulation Service* (CES). This service emulates a point-to-point T1 or E1 circuit and a point-to-point fractional T1 or E1 circuit over an ATM network. The fractional T1 or E1 service provides a N $\times$ 64 Kbps circuit, where *N* can be as small as 1 and as high as 24 for T1 and 31 for E1. CES is connection-oriented, and the user equipment attached at each end of the emulated circuit transmits at a constant bit rate. Clock synchronization of the user equipment can be maintained by passing timing information over the ATM network.

Examples of class B service are audio and video services transmitted at a variable bit rate. These are connection-oriented services, where the user equipment transmit at a variable bit rate and require clock synchronization. As in class A, clock synchronization can be achieved by passing timing information from the sender to the receiver over the ATM network.

Service classes C and D do not provide clock synchronization between sender and receiver. A frame relay service provides connection-oriented data transfers at a variable transmission rate, and it is a good candidate for Class C. On the other hand, an application running on top of TCP/IP can use class D.

In addition to these four classes, class X was defined as the *raw cell* service. It allows the use of a proprietary AAL with terminal equipment that supports a vendor-defined AAL. That is, the AAL, traffic type (i.e. variable or constant bit rate) and timing requirements are user-defined.

Corresponding to these four service classes are four ATM adaptation layers, referred to as *AAL 1*, *AAL 2*, *AAL 3/4* and *AAL 5*. These adaptation layers are used for the transfer of user data, and they are described in detail in this chapter. A fifth adaptation layer used by the ATM signaling protocols, the Signaling ATM Adaptation Layer (SAAL), is described in Chapter 10.

5.2 ATM ADAPTATION LAYER 1 (AAL 1)

This AAL was proposed for class A services, and it can be used for applications such as circuit emulation services, constant-bit rate video, and high-quality constant-bit rate audio. It provides transfer of constant-bit rate data, delivery at the same bit rate, and transfer of timing information between the sending and receiving applications. Also, it can handle cell delay variation, and can detect lost or misrouted cells.

AAL 1 consists of a SAR sublayer and a Convergence Sublayer (CS). The SAR sublayer is responsible for the transport and bit error detection, and possibly correction, of blocks of data received from CS. The CS sublayer performs a variety of functions, such as handling the cell delay variation, processing the sequence count, structured and unstructured data transfers, and transfer of timing information.

5.2.1 The AAL 1 SAR sublayer

The SAR sublayer accepts blocks of 47 bytes from the convergence sublayer, and adds a 1-byte header to form the SAR-PDU. The SAR-PDU is then passed on to the ATM layer, where it gets encapsulated with a 5-byte ATM header. The ATM cell is then passed on to the physical layer, which transmits it out. At the receiving SAR sublayer, the 1-byte header is stripped and the payload of the SAR-PDU is delivered to the receiving CS.

The encapsulation of the SAR PDU is shown in Figure 5.3. The header consists of two fields, the Sequence Number (SN) field and the Sequence Number Protection (SNP) field. Both fields are four bits long.

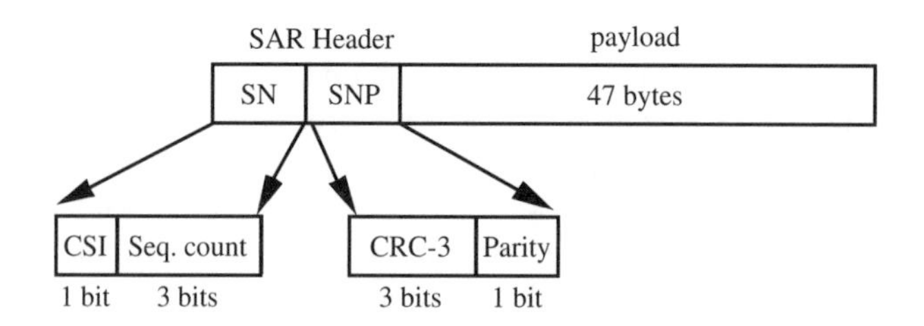

Figure 5.3 The SAR encapsulation for AAL 1.

The SN field contains two subfields:

- *Convergence Sublayer Indication* (CSI): carries an indication which is provided by the CS. The default value of the CSI bit is '0'.
- *Sequence count*: this is provided by the transmitting CS, and it is associated with the block of data in the SAR-PDU. The count starts at 0 and is increased sequentially modulo 8. It is used by the receiving CS to detect lost or misinserted cells.

The SNP field contains the following two subfields:

- *CRC-3*: it is computed over the CSI and sequence count fields.
- *Parity*: even parity bit used calculated over the CSI, sequence count, and CRC-3 fields.

The transmitting SAR computes the FCS for the first four bits of the header and inserts it in the CRC-3 field. The pattern used to compute the FCS is given by the polynomial: $x^3 + x + 1$. After completing the CRC operation, the transmitting AAL calculates the even parity bit on the first seven bits of the header, and inserts the result in the parity field.

The receiving SAR examines each SAR-PDU header by checking the FCS and the even parity bit. The state machine that controls the receiver's error detection and correction scheme is the same as the header error control scheme used for the ATM header, described in Section 4.2 and shown in Figure 4.4. At initialization, the state machine is set to the correction mode. Each time a SAR-PDU comes in, the FCS and the parity bit are checked, and if no errors are found, the SN field is declared to be valid and the state machine remains in the correction mode. If a single-bit error is detected, it is corrected and the SN field is declared to be valid, but the state machine switches to detection mode. If a multi-bit error is detected, the SN field is declared to be invalid and the state machine switches to detection mode. In detection mode, each time a SAR-PDU comes in, the FCS and the parity bit are checked and if a single-bit or a multi-bit error is detected, the SN field is declared to be invalid and the state machine remains in detection mode. If no errors are detected, then the SN field is declared to be valid and the state machine shifts back to the correction mode.

The receiving SAR sublayer conveys to the receiving CS, the sequence count, the CS indication, and the status of the SN field, i.e. valid or invalid.

This error detection and correction scheme runs in addition to the error detection and correction scheme for the ATM header. However, these two mechanisms apply to two different fields of the ATM cell. The header error control mechanism applies to the first four bytes of cell's header, whereas the above scheme applies to the SN field.

5.2.2 The AAL1 CS sublayer

The convergence sublayer performs various functions, such as handling of the cell delay variation, processing of the sequence count, forward error correction, performance monitoring, structured and unstructured data transfers, and transfer of timing information. Below, we describe each of these functions.

Handling of the cell delay variation

For AAL1 to support constant bit rate applications, it has to deliver the data stream to the receiving application at the same bit rate at which it was transmitted. The data stream, as we have seen, is carried by ATM cells which may have to traverse a number of ATM switches before they reach their destination. In view of this, the arrival of these cells may be occasionally delayed because of congestion in the network. Also, the opposite may occur. That is, a group of cells may arrive closer to each other than they were originally transmitted.

To compensate for this variability in the arrival of the cells, CS writes the incoming SAR-PDUs into a buffer, from where it delivers the data stream to the receiving AAL application at a constant bit rate. (A similar method is used, for instance, when we listen to a radio station over the Internet.) In the event of buffer underflow, it may be necessary for the CS to insert dummy bits in order to maintain bit count integrity. Also, in the event of buffer overflow, CS drops the appropriate number of bits.

Processing of the sequence count

The sequence count values are processed by CS to detect lost or misinserted cells. Detected misinserted cells are discarded. To maintain bit count integrity of the AAL user information, it may be necessary to compensate for lost cells by inserting dummy SAR-PDU payloads.

Forward error correction

For video and high quality audio, forward error correction may be performed to protect against bit errors. This may be combined with interleaving of AAL user bits to give a more secure protection against errors.

Performance monitoring

The CS may generate reports giving the status of end-to-end performance, as deduced by the AAL. The performance measures in these reports may be based on events of lost or misinserted cells, buffer underflows and overflows, and bit error events.

Structured and unstructured data transfers

We first note that two CS-PDU formats have been defined to support structured and unstructured data transfers. These two formats are the *CS-PDU non-P format* and the *CS-PDU P format*. The CS-PDU non-P format is constructed from 47 bytes of information supplied by an AAL user. The CS-PDU P format is constructed from a 1-byte header and

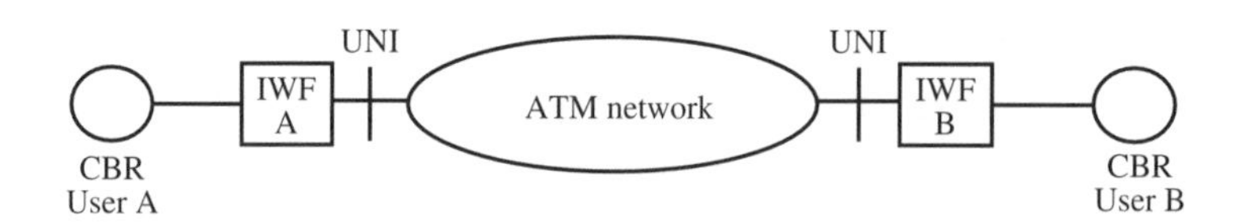

Figure 5.4 The interworking function in CES.

46 bytes of information supplied by the AAL user. The 1-byte header contains a single field, the *Structured Pointer* (SP), used to point to a block of data. The CS-PDU P format, or non-P format, becomes the payload of the SAR-PDU.

Structured and unstructured data transfers are used in Circuit Emulation Services (CES), which emulate a T1/E1 link over an ATM network using AAL 1. To implement CES an *InterWorking Function* (IWF) is required, as shown in Figure 5.4. The two IWFs are connected over the ATM network using AAL 1 via a bi-directional point-to-point virtual circuit connection. CES provides two different services, namely *DS1/E1 unstructured service* and *DS1/E1 Nx64 Kbps structured service*.

In the unstructured service, CES simply carries an entire DS1/E1 signal over an ATM network. Specifically, IWF A receives the DS-1 signal from user A which it packs bit-by-bit into the 47-byte non-P format CS-PDU, which in turn becomes the payload of a SAR-PDU. The SAR-PDUs then become the payload of ATM cells which are transferred to IWF B, from where the bits get delivered to B as a DS-1 signal.

In the structured service, CES provides a fractional DS1/E1 service, where the user only requires an Nx64 Kbps connection. N can be as small as 1 and as high as 24 for T1 and 30 for E1. An Nx64 Kbps service generates blocks of N bytes, which are carried in P and non-P format CS-PDUs. The beginning of the block is pointed to by the pointer in the 1-byte header of the CS-PDU P format.

Transfer of timing information

Some AAL users require that the clock frequency at the source is transferred to the destination. CS provides mechanisms for transferring such timing information.

If the sender's clock and the receiver's clock are phase-locked to a network's clock, then there is no need to transfer the source's clock frequency to the receiver, and AAL 1 is not required to transfer any timing information. However, if the two clocks are not phase-locked to a network's clock, then AAL 1 is required to transfer the source clock frequency to the receiver. This can be done using the *Synchronous Residual Time Stamp* (SRTS) method, where AAL 1 conveys to the receiver the difference between a common reference clock derived from the network and the sender's service clock. This information is transported in the CSI bit of the SAR-PDU header over successive cells. The common reference clock has to be available to both the sender and receiver. This is the case, for instance, when they are both attached to a synchronous network like SONET.

When a common reference clock is not available, the *adaptive clock method* can be used. In this method, the receiver writes the received information into a buffer and then reads out from the buffer with a local clock. The fill level of the buffer is used to control the frequency of the local clock as follows. The buffer fill is continuously measured, and if it is greater than the median, then the local clock is assumed to be slow and its speed is increased. If the fill level is lower than the median, then the clock is assumed to be fast and its speed is decreased.

5.3 ATM ADAPTATION LAYER 2 (AAL 2)

This adaptation layer was originally defined for class B services. It was later redefined to provide an efficient transport over ATM for multiple applications which are delay-sensitive and have a low variable bit rate, such as voice, fax and voiceband data traffic. Specifically, AAL 2 was designed to multiplex a number of such low variable bit rate data streams on to a single ATM connection. At the receiving side, it demultiplexes them back to the individual data streams. An example of AAL 2 is given in Figure 5.5. In this example, the transmitting AAL 2 multiplexes the data streams from users A, B, C and D, onto the same ATM connection. The receiving AAL 2 demultiplexes the data stream into individual streams and delivers each stream to the peer user A′, B′, C′ and D′.

Target networks for AAL2 are narrowband and wideband cellular networks, and private narrowband networks (PBX).

The AAL 2 services are provided by the convergence sublayer, which is subdivided into the Service Specific Convergence Sublayer (SSCS) and the Common Part Sublayer (CPS). There is no SAR layer in AAL 2. The multiplexing of the different user data streams is achieved by associating each user with a different SSCS. Different SSCS protocols may be defined to support different types of service. Also, the SSCS may be null. Each SSCS receives data from its user, and passes this data to the CPS in the form of short packets. The CPS provides a multiplexing function, whereby the packets received from different SSCS are all multiplexed onto a single ATM connection. At the receiving side of the ATM connection, these packets are retrieved from the incoming ATM cells by CPS and delivered to their corresponding SSCS receivers. Finally, each SSCS delivers its data to its user. The functional model of AAL 2 at the sender's side is shown in Figure 5.6

A transmitting SSCS typically uses a timer to decide when to pass on data to CPS. When the timer expires, it passes the data that it has received from its higher-layer application to CPS in the form of a packet, known as the *CPS-packet*. Since the applications that use AAL 2 are low variable bit rate, the CPS-packets are very short and they may have a variable length. Each CPS-packet is encapsulated by CPS and then it is packed into a *CPS-PDU*. As mentioned above, AAL 2 has been designed to multiplex several SSCS streams onto a single ATM connection. This is done by packing several CPS-packets

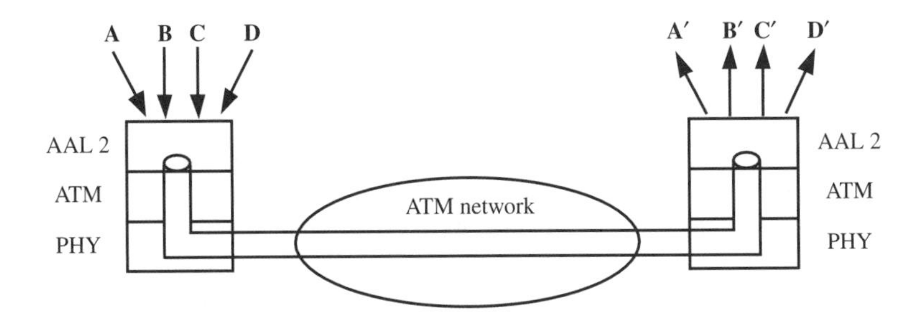

Figure 5.5 AAL 2 can multiplex several data streams.

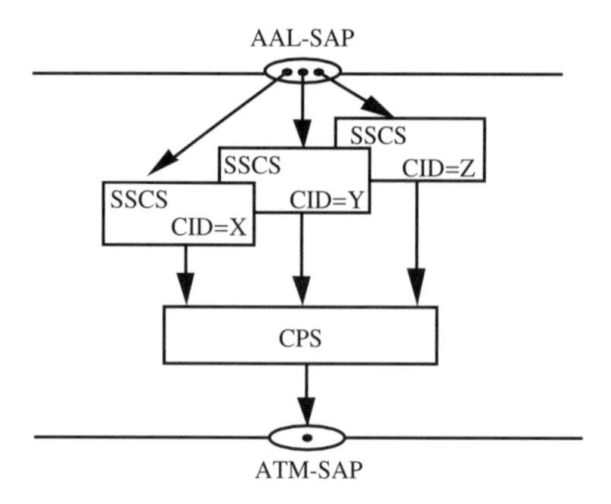

Figure 5.6 Functional model of AAL 2 (sender side).

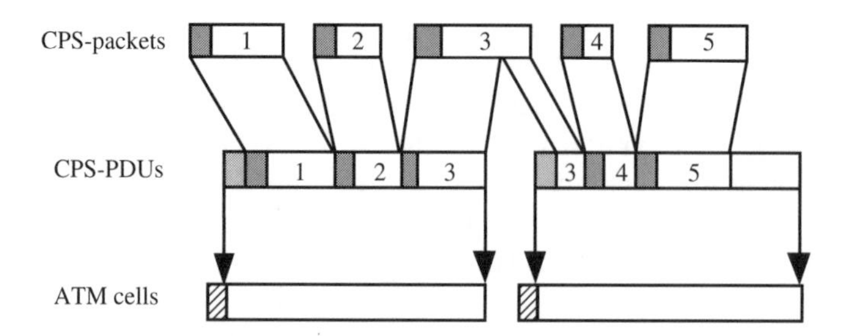

Figure 5.7 Packing CPS-packets into CPS-PDUs.

into a single CPS-PDU, where each CPS-packet belongs to a different SSCS stream. It is possible for a CPS-packet to straddle over two successive CPS-PDUs, as shown in Figure 5.7. In this figure, we see that CPS-packets 1 and 2 fit entirely in a CPS-PDU, whereas CPS-packet 3 has to be split between the first and the second CPS-PDU. The point where a CPS-packet is split may occur anywhere in the CPS-packet, including the CPS-packet header. The unused payload in a CPS-PDU is padded with zero bytes.

CPS-packets are encapsulated with a three-byte header, and CPS-PDUs with a one-byte header. The CPS-PDU, after encapsulation, is exactly 48 bytes long, and it becomes the payload of an ATM cell. The structure of the CPS-packet and the CPS-PDU is shown in Figure 5.8.

The header of the CPS-packet contains the following fields:

- *Channel identifier* (CID): CPS can multiplex several streams, referred to as *channels*, onto a single ATM connection. Each channel is identified by the channel identifier, given in the 8-bit field CID. A channel is bidirectional and the same CID value is used in both directions. CID values are allocated as follows: 0 is not used as a channel identification because it is used as padding; 1 is used by *AAL 2 Negotiation Procedure* (ANP) packets, described below; 2 to 7 are reserved; and 8 to 255 are valid CID values used to identify channels.

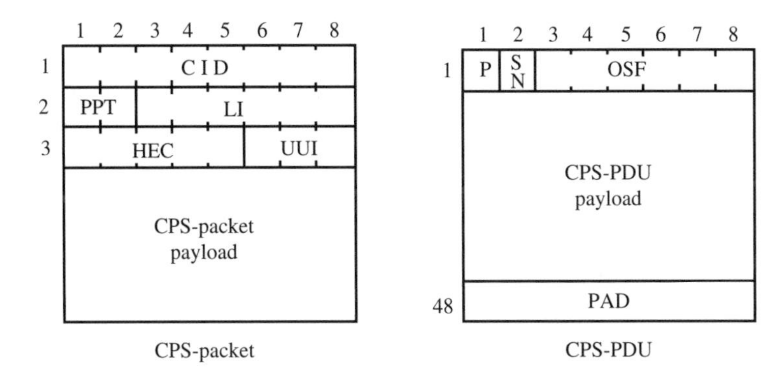

Figure 5.8 The structure of the CPS-packet and CPS-PDU.

- *Packet Payload Type* (PPT): the PPT is a 2-bit field and it serves two types of functions. When PPT $\neq 3$, the CPS-packet is serving a specific application, such as carrying voice data, or carrying an ANP packet. When PPT $= 3$, the CPS-packet is serving an AAL network management function associated with the management of the channel identified in the CID field.
- *Length Indicator* (LI): the total number of bytes in the payload of the CPS-packet is indicated in this 6-bit field. Its value is one less than the number of bytes in the CPS-packet payload. The default maximum length of the CPS-packet payload is 45 bytes. A maximum length of 64 bytes can be negotiated by ANP or by the network management protocol.
- *Header Error Control* (HEC): this 5-bit field carries the FCS obtained from the CRC operation carried over the CID, PPT, LI and UUI fields using the pattern $x^5 + x^2 + 1$. The receiver uses the contents of the HEC to detect errors in the header.
- *User-to-User-Indication* (UUI): this is a 3-bit field, and it is used to transfer information between the peer CPS users. This information is transported by the CPS transparently.

The header of the CPS-PDU is referred to as the *start field* (STF), and it contains the following fields:

- *Parity* (P): a 1-bit field used to detect errors in the STF. The transmitter sets this field so that the parity over the 8-bit STF is odd.
- *Sequence Numbers* (SN): a 1-bit field used to number modulo 2 the successive CPS-PDUs.
- *Offset field* (OSF): the CPS-PDU payload may carry CPS packets in a variety of different arrangements. For instance, in the example given in Figure 5.7, the first CPS-PDU contains two complete CPS-packets, CPS-packet 1 and 2, followed by a partial CPS-packet, CPS-packet 3. The second CPS-PDU in the same figure consists of the remaining of CPS-packet 3, two complete packets, CPS-packets 4 and 5, and padding.

To help the receiving CPS extract the CPS-packets from the payload of a CPS-PDU, a 6-bit offset field (OSF) is used to indicate the start of a new CPS-packet in the CPS-PDU payload. Specifically, OSF gives the number of bytes between the end of the STF (i.e. the header of the CPS-PDU) and the start of the first CPS-packet in the CPS-PDU payload.

In the absence of a start of a CPS-packet, it gives the start of the PAD field. The value 47 indicates that there is no beginning of a CPS-packet in the CPS-PDU.

We note that the AAL 2 CPS transfers the CPS-PDUs in a nonassured manner. That is, a CPS-PDU may be delivered or it may be lost. Lost CPS-PDUs are not recovered by retransmission.

The function that provides the dynamic allocation of AAL 2 channels on demand is called the *AAL Negotiation Procedures* (ANP). This function is carried out by an AAL 2 layer management entity at each side of an AAL 2 link. This layer management entity uses the services provided by AAL 2 through a SAP for the purpose of transmitting and receiving ANP messages. These messages are carried on a dedicated AAL 2 channel with CID = 1, and they control the assignment, removal and status of an AAL 2 channel. The following types of messages have been defined: assignment request, assignment confirm, assignment denied, removal request, removal confirm, status poll, and status response.

5.4 ATM ADAPTATION LAYER 3/4 (AAL 3/4)

This adaptation layer was first designed to be used in SMDS (Switched Multi-Megabit Data Service), and later it was standardized for ATM networks. It was originally proposed for class C services, but very quickly it became obvious that it can also be used for class D services. In view of this, it was named AAL 3/4. This adaptation layer achieves per-cell integrity, as it can detect a variety of different errors related to the segmentation and reassembly of packets. However, it is quite complex and has a lot of overhead. AAL 3/4 was originally favored by the telecommunications industry, but due to its complexity, it was not very popular within the computer industry. In reaction to AAL 3/4, a simpler adaptation layer, known as AAL 5, was proposed. AAL 5 is a very popular ATM adaptation layer, and is presented in Section 5.5.

The AAL 3/4 functions are provided by the convergence sublayer and the SAR sublayer. The convergence sublayer is subdivided into the Service Specific Convergence Sublayer (SSCS) and the Common Part Sublayer (CPS). We note that in the standards CPS is referred to as the *Common Part Convergence Sublayer* (CPCS).

Different SSCS protocols may be defined to support specific AAL 3/4 applications. SSCS may also be null, which is the assumption made in this section. AAL 3/4 provides a *message mode* service and a *streaming mode* service. In message mode, a user-PDU is transported in a single CPS-PDU. It may also be segmented, and each segment is transported in a single CPS-PDU. In streaming mode, one or more user-PDUs are blocked together and transported in a single CPS-PDU.

Both message mode and streaming mode may offer an *assured* or a *nonassured* peer-to-peer operation. In the assured operation, every CPS-PDU is delivered with exactly the data content that the user sent. The assured service is provided by retransmitting missing or corrupted CPS-PDUs. Flow control may be used optionally. The assured operation may be restricted to point-to-point ATM connections. In the nonassured operation, lost or corrupted PDUs are not corrected by retransmission. An optional feature may be provided to allow corrupted PDUs to be delivered to the user. Flow control may be provided as an option for point-to-point ATM connections.

The encapsulation at the transmitting CPS and SAR is shown in Figure 5.9. CPS adds a CPS-PDU header and a CPS-PDU trailer for each user-PDU, to form a CPS-PDU. The CPS-PDU is then segmented, and each segment is encapsulated with a SAR-PDU header

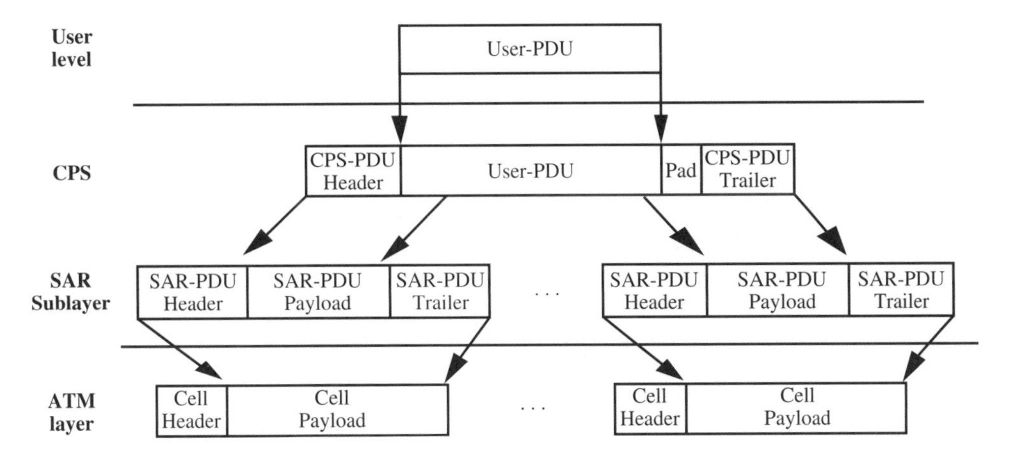

Figure 5.9 The CPS and SAR headers and trailers (sender side).

and a SAR-PDU trailer to form a SAR-PDU. Each SAR-PDU becomes the payload of an ATM cell.

The SAR layer introduces a 2-byte header and a 2-byte trailer as shown in Figure 5.10. The following fields have been defined:

- *Segment Type* (ST): four different segment types are defined, namely, *Beginning Of Message* (BOM), *Continuation Of Message* (COM), *End Of Message* (EOM) and *Single Segment Message* (SSM). The coding of the message type is as follows: 10 (BOM), 00 (COM), 01 (EOM), 11 (SSM). The segment type is used to indicate the position of a segment within a CPS-PDU.
- *Sequence Number* (SN): all the SAR-PDUs associated with the same CPS-PDU are numbered sequentially. This number is carried in the sequence number field. The sender can select any value between 0 and 15 as the sequence number of the first SAR-PDU of a CPS-PDU. These sequence numbers do not continue on to the next CPS-PDU.
- *Multiplexing Identifier* (MID): multiple connections may be multiplexed over the same ATM connection. Each of these multiplexed connections is identified by a different MID number. If no multiplexing is used, this field is set to zero.
- *Length Identification* (LI): gives the number of bytes of CPS-PDU that are included in the SAR-PDU (value ≤ 44).

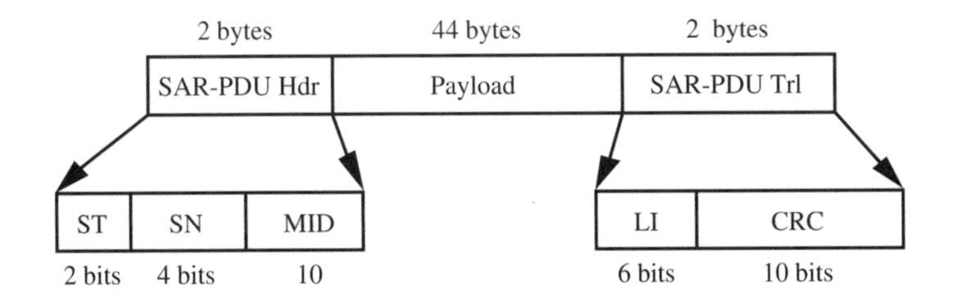

Figure 5.10 SAR encapsulation.

- *CRC*: this field contains the FCS obtained by carrying out a cyclic redundancy check over the entire contents of the SAR-PDU using the pattern $x^{10} + x^9 + x^5 + x^4 + x + 1$.

CPS introduces a 4-byte header and a 4-byte trailer as shown in Figure 5.11. The following fields have been defined:

- *Common Part Indicator* (CPI): identifies the convergence service to be provided. *Beginning/End tag* (Btag/Etag). It has the same value in the header Btag and trailer Etag. This field permits us to check the association of the CPS-PDU header and trailer, and it changes for each successive CPS-PDU.
- *Buffer Allocation size* (BA-size): used to indicate to the receiver the maximum buffer requirements necessary to receive the CPS-PDU.
- *Pad*: this field is used so that the CPS payload is an integer multiple of 32 bits. The pad maybe from 0 to 3 bytes, and it does not convey any information.
- *Alignment field* (AL): this field is used in order to maintain a 32 bit alignment of the CPS PDU trailer. It is an unused byte, and it is strictly used as filler.
- *Length*: specifies in bytes the length of the user-PDU.

The reassembly of the ATM cells into the original user-PDUs is shown in Figure 5.12. The SAR-PDUs are extracted from the ATM cells, stripped of their headers and trailers,

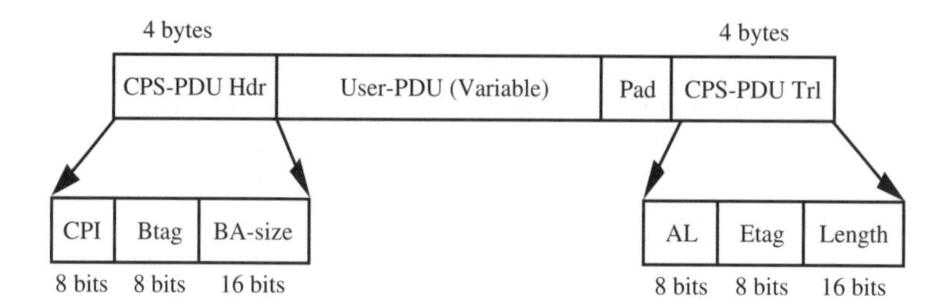

Figure 5.11 CPS encapsulation.

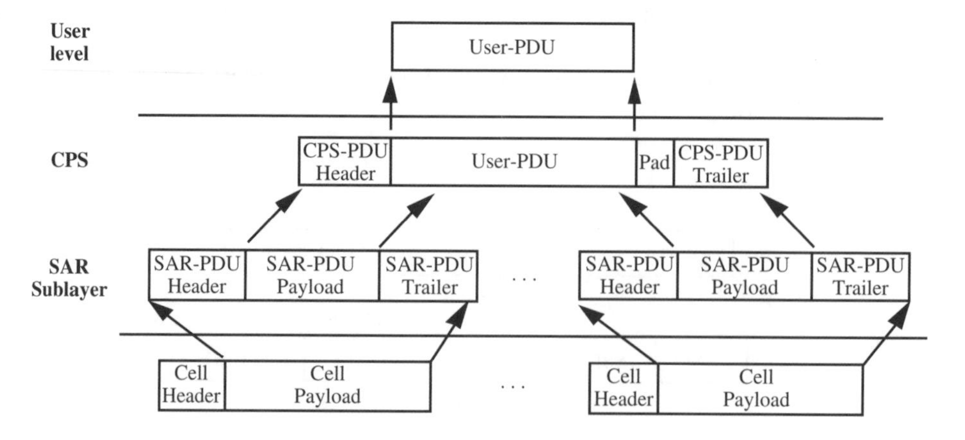

Figure 5.12 Reassembly at the receiver.

and are reassembled into a CPS-PDU. The CS-PDU is stripped off its header and trailer, and is delivered to the higher layer.

5.5 ATM ADAPTATION LAYER 5 (AAL 5)

This adaptation layer is suitable for connectionless services (class D). It is also used in the Signaling AAL (SAAL) described in Chapter 10. It was proposed in reaction to AAL 3/4, which is unnecessarily complex.

The AAL 5 services are provided by the Convergence Sublayer (CS) and the SAR sublayer. CS is further subdivided into the SSCS and CPS. We note that in the standards CPS is referred to as the Common Part Convergence Sublayer (CPCS).

Different SSCS protocols may be defined to support specific AAL users. The SSCS may be also null, which is the assumption made in this section.

CPS provides a nonassured transfer of user-PDUs with any length that can vary from 1 byte to 65 535 bytes. At the receiving side, CPS can detect erroneous CPS-PDUs, and it can indicate that to the higher-level application. However, since it provides a nonassured service, it does not recover erroneous CPS-PDUs by retransmission. This is left to a higher-level protocol, such as TCP. It also delivers user-PDUs in the order in which it received them.

CPS provides both message mode and streaming mode services. In message mode, it is passed a user-PDU which it transfers in a single CPS-PDU. In streaming mode, it is passed several user-PDUs over a period of time, which it blocks together and transports to the destination in a single CPS-PDU.

A user-PDU is encapsulated by CPS into a CPS-PDU by adding a trailer, as shown in Figure 5.13. The following fields in the trailer have been defined:

- *Padding* (Pad): can be between 0 and 47 bytes long, and is added so that the entire CPS-PDU including the padding and the remaining fields in the trailer becomes an integer multiple of 48 bytes. The padding is made up of unused bytes which do not convey any information.
- *CPS User-to-User indication* (CPS-UU): a 1-byte field used to transfer transparently CPS user-to-user information.
- *Common Part Indicator* (CPI): a 1-byte field to support future AAL 5 functions.
- *Length*: a 2-byte field used to indicate the length in bytes of the CPS-PDU payload, i.e. the user-PDU. This is used by the receiver to detect if there has been loss or gain of information.
- *CRC-32*: this 4-byte field contains the FCS calculated by the transmitting CPS over the entire contents of the CPS-PDU, i.e. the user-PDU, pad, CPS-UU, CPI and length. The pattern used is: $x^{32} + x^{26} + x^{23} + x^{22} + x^{16} + x^{12} + x^{11} + x^{10} + x^8 + x^7 + x^5 + x^4 + x^2 + x + 1$.

The SAR sublayer fragments a CPS-PDU into a sequence of 48-byte segments, and each segment is carried in the payload of an ATM cell. There is no encapsulation at

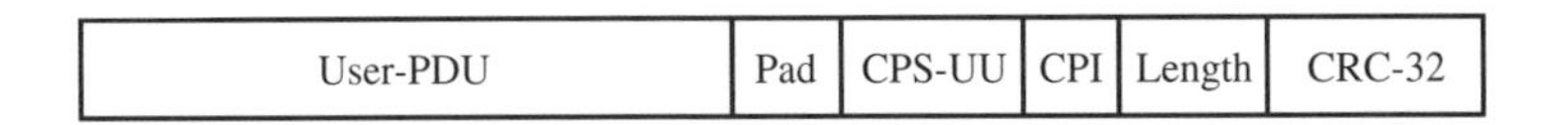

Figure 5.13 Encapsulation of a user-PDU.

the SAR sublayer, as in AAL 3/4. However, the ATM cell that carries the last segment of a CPS-PDU is marked by setting the SDU type indication in its PTI field to 1 (see Table 4.2). Specifically, each ATM cell that carries a segment of a CPS-PDU has its SDU type indication set to zero, except the ATM cell that carries the last segment of the CPS-PDU whose PTI field contains the indication SDU type = 1.

The receiving SAR appends the payloads of the ATM cells into a buffer, associated with the VCC over which the ATM cells are being transmitted, until it either encounters an ATM cell with an indication SDU-type = 1 in its PTI field, or the CPS-PDU exceeds the buffer. Upon occurrence of either event, the buffer is passed on to a higher-level application with an indication as to whether the indication SDU-type = 1 was received or not, and in the case it was received, whether the CRC check was correct or not.

PROBLEMS

1. Consider the case where a DS-1 signal is transported over ATM via AAL 1 using the unstructured data transfer mode. How long does it take to fill-up a SAR-PDU?
2. In AAL 2, the receiving CPS retrieves the CPS-packets carried in the payload of a CPS-PDU using the offset field (OSF) in the header of the CPS-PDU and the LI field in the header of each CPS-packet carried in the CPS-PDU. The use of the OSF may appear redundant! Construct an example where the receiving CPS cannot retrieve the CPS-packets carried in a CPS-PDU by using only the LI field in each CPS-packet. Assume no cell loss or corrupted ATM payloads.
3. In AAL 2, why does the value 0 cannot be used to indicate a CID?
4. The following CPS-packets have to be transmitted over the same AAL2 connection: CPS-packet 1 (20 bytes), CPS-packet 2 (48 bytes), CPS-packet 3 (35 bytes), and CPS-packet 4 (20 bytes). For simplicity, assume that the length of each of these CPS-packets includes the three-byte CPS-packet header.
   (a) How many CPS-PDUs are required to carry these four CPS-packets?
   (b) What is the value of the OSF field in each CPS-PDU?
5. A voice source is active (talkspurt) for 400 ms and silent for 600 ms. Let us assume that a voice call is transported over an ATM network via AAL 2. The voice is coded to 32 Kbps and silent periods are suppressed. We assume that the SSCS has a timer set to 5 ms. That is, each time the timer expires, it sends the data it has gathered to CPS as a CPS-packet. Assume that the timer begins at the beginning of the busy period.
   (a) How long (in bytes) is each CPS-packet?
   (b) How many CPS-packets are produced in each active period?
6. A 1500-byte user-PDU is transported over AAL3/4 in a single CPS-PDU.
   (a) How many additional bytes are introduced by CPS?
   (b) How many SAR-PDUs are required to carry the resulting CPS-PDU?
   (c) How many additional bytes are introduced by the SAR in each SAR-PDU?
   (d) What is the maximum and minimum useful payload in each ATM cell that carry the resulting SAR-PDUs (i.e. the maximum and minimum of the number of bytes belonging to the original user-PDU)?
7. A 1500-byte user-PDU is transported over AAL5.
   (a) How many bytes of padding will be added?
   (b) How many cells are required to carry the resulting CPS-PDU?
   (c) What is the total overhead (i.e. additional bytes) associated with this user-PDU?

6

ATM Switch Architectures

The advent of ATM gave rise to a new generation of switches capable of switching cells at high speeds. These ATM switches can be grouped into three classes, *space-division switches, shared memory switches* and *shared medium switches*. In this chapter, we describe these three types of architectures in detail. Also, we present various scheduling policies that can be used to schedule the transmission of cells out of a switch. Finally, we describe in some detail an existing switch.

6.1 INTRODUCTION

The main function of an ATM switch is to transfer cells from its incoming links to its outgoing links. This is known as the *switching* function. In addition, a switch performs several other functions, such as signaling and network management. A generic model of an ATM switch consisting of N input ports and N output ports is shown in Figure 6.1. Each input port may have a finite capacity buffer where cells wait until they are transferred to their destination output ports. The input ports are connected to the output ports via the *switch fabric*. Each output port may also be associated with a finite capacity buffer, where cells can wait until they are transmitted out. Depending upon the structure of the switch fabric, there may be additional buffering inside the fabric. An ATM switch whose input ports are equipped with buffers, irrespective of whether its output ports have buffers or not, is referred to as an *input buffering* switch. An ATM switch is referred to as an *output buffering* switch if only its output ports are equipped with buffers. Depending on the switch architecture, cell loss may occur at the input ports, within the switch fabric, and at the output ports.

An ATM switch is equipped with a CPU, which is used to carry out signaling and management functions.

We note that the switch in Figure 6.1 was drawn as an *unfolded* switch, i.e. the input and output ports were drawn separately with the switch fabric in-between. As a result, the flow of cells from the input ports to the output ports is from left to right. In a real-life switch, each port is typically *dual*, i.e. it is both an input and an output port. The link connecting to a port may be either duplex (i.e. it can transmit in both directions at the same time), or it may consist of two separate cables, one for transmitting cells to the switch and the other for transmitting cells out of the switch. In Figure 6.2, we show an ATM switch drawn as a *folded* switch, i.e. each port is shown both as an input and an output port. For instance, user A is attached to port 1, which means that it transmits cells to the switch and receives cells from the switch via port 1. User B is attached to port N, and therefore it transmits cells to the switch and receives cells from the switch via port

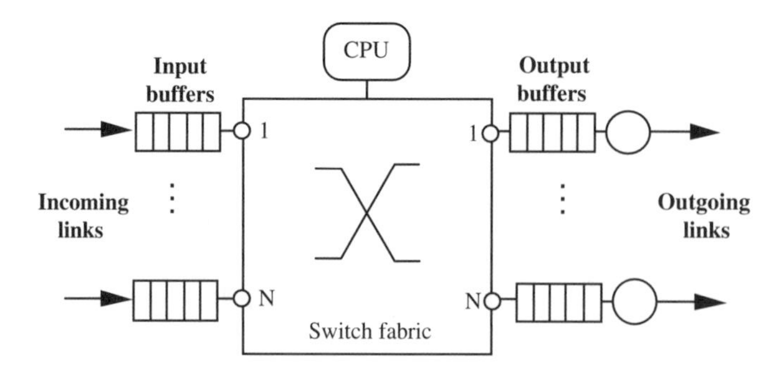

Figure 6.1 A generic model of an ATM switch.

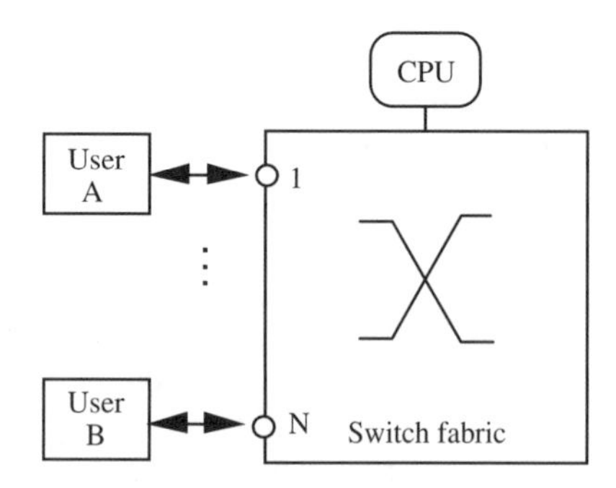

Figure 6.2 A folded ATM switch.

N. For presentation purposes, switches are typically drawn unfolded, and in this book we follow the same convention. However, it is important to remember that when referring to a diagram of an unfolded switch, the device that transmits to the switch on input port i, also receives cells from the switch from output port i.

Header conversion typically takes place before a cell is transferred to its destination output port. The value of the VPI/VCI fields of an arriving cell is looked up in a table, which provides the new VPI/VCI values and the destination output port number. Typically, this table is not very large. In view of this, the table look-up function is not a time-consuming process. Depending upon the type of switch architecture and the way in which it has been implemented, there may be a single header conversion table for the entire switch serving all the input ports, or one header conversion table per input port. An example of such a table assuming a 16×16 switch is shown in Table 6.1.

An ATM switch can be deployed at the edge of an ATM network or inside an ATM network. A switch deployed at the edge of an ATM network may be equipped with various interfaces, such as ADSL, FDDI, Ethernet and token ring, so that it can receive and transmit traffic from/to residential access networks, LANs and MANs. A switch deployed inside an ATM network has only ATM interfaces.

ATM switch architectures can be grouped into the following three classes: *space-division switch, memory sharing switch* and *medium sharing switch*.

Table 6.1 A header conversion table.

VPI	VCI	Input port	VPI	VCI	Output port
10	20	1	12	43	12
11	21	16	13	44	10
12	20	10	14	44	8
43	12	12	14	43	2

Space-division switch architectures are based on *Multi-stage Interconnection Networks* (MIN). A MIN consists of a network of interconnected switching elements, arranged in rows and columns. MINs were originally used to construct telephone switches. Later, they were used in tightly coupled multiprocessor systems to interconnect processors to memory modules, and in the early 1990s they were used extensively to construct ATM switches.

A shared memory switch architecture utilizes a single memory which is used to store all the cells arriving from the input ports. Cells stored in the memory are organized into linked lists, one per output port. The cells in each linked list are transmitted out of the switch by its associated output port. ATM shared memory switches are currently very popular.

Finally, in a medium sharing switch, all arriving cells at the switch are synchronously transmitted onto a bus. Each output port i can see all the cells transmitted on the bus, and it receives those cells whose destination is output port i. There is a buffer in front of each output port, where the cells can wait until they are transmitted out.

In the following sections, we describe in detail each of these three types of switch architectures.

6.2 SPACE-DIVISION SWITCH ARCHITECTURES

Space-division ATM switch architectures have been extensively studied. The main features of these architectures is that control of the switching function is not centralized. Also, multiple concurrent paths can be established from input ports to output ports, each with the same data rate as an incoming link. We will examine the following switching fabrics: *cross-bar*, *banyan*, *Clos* and N^2 *disjoint paths*.

Depending on the structure of the switching fabric, it may not be possible to setup the required paths from the input ports to the output ports without any conflicts, i.e. two (or more) paths requiring the same link at the same slot. Blocking the transfer of a cell from its input port to its output port, because a link along its path is used by another cell following a different path, is known as *internal blocking*. Another type of blocking, which is referred to as *external blocking*, occurs when two or more cells seek the same output port at the same time.

6.2.1 The Cross-Bar Switch

This is a very popular fabric, and has been used extensively in circuit switching and packet switching. It consists of a square array of $N \times N$ crosspoint switches, as shown in

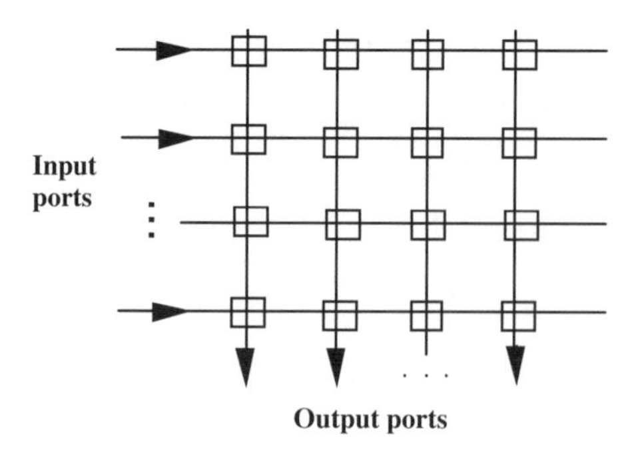

Figure 6.3 A cross-bar switch.

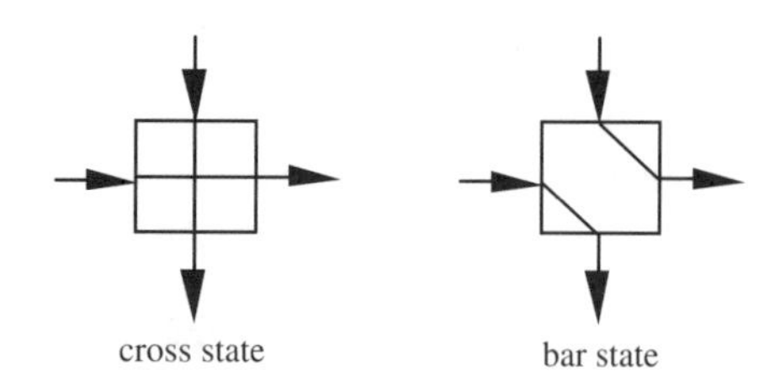

Figure 6.4 The cross state and bar state of a crosspoint.

Figure 6.3. Each crosspoint can assume a cross state or a bar state, as shown in Figure 6.4. In the cross state, the horizontal input is connected to the horizontal output, and the vertical input is connected to the vertical output. In the bar state, the horizontal input is connected to the vertical output, and the vertical output is connected to the horizontal output. All crosspoints are originally in the cross state.

Let us assume that two cells arrive at a crosspoint at the same time, one on the horizontal input and the other on the vertical input. Let us also assume that the crosspoint is in the cross state. Then, both cells can successfully go through the crosspoint if the destination output of the cell on the horizontal input is the horizontal output, and the destination output of the cell on the vertical input is the vertical output. However, if both cells want to go out of the same output, say the vertical output, then only the cell coming in on the vertical input will go through. A similar operation takes place when the crosspoint is in the bar state.

We shall refer to a crosspoint by its row and column number. For instance, crosspoint (i,j) lies at the intersection of the ith row and the jth column, where the ith row is associated with the ith input port and the jth column is associated with the jth output port.

The switch fabric is self-routing. A cell at input port i destined for output port j is first propagated horizontally from input port i to the crosspoint (i,j) by setting all the crosspoints on its way to the cross state. When it gets to crosspoint (i,j), it sets it to the bar state, and then proceeds to traverse vertically downwards to the output port j by also setting the crosspoints on its way to the cross state. This is the only path that a cell can follow from input port i to output port j. No other paths through the switch fabric are feasible. For instance, it is not possible for the cell to be propagated horizontally across a few crosspoints, then vertically across a few more crosspoints, then again horizontally across a few more crosspoints, and so on, until it gets to its destination output port.

Only one cell per input port can be launched into the switch fabric at a time. Thus, in an $N \times N$ switch fabric, at most N cells can be transferred to their output ports at the same time. This can only occur when each cell has a different destination output port. For example, let us assume that a cell from input port 1 is destined to output port 5, and a cell from input port 2 is destined to output port 3. Then, both cells can be transmitted at the same time without any conflict, since they use different links of the cross-bar switch fabric. Specifically, the cell from input port 1 will be propagated horizontally until it reaches crosspoint (1,5), it will set it to the bar state, and will then traverse vertically down to output port 5. Likewise, the cell from input port 2 will traverse horizontally until it reaches crosspoint (2,3). It will set it to the bar state and then traverse vertically to output port 3. Now, let us assume that both cells have the same destination output port. In this case, conflict will arise as the two cells eventually will have to use the same vertical links. This is an example of internal blocking. It can also be seen as external blocking, since the two cells compete for the same output port. In fact, the distinction between internal and external blocking is not very clear in this switch fabric.

The cross-bar switch is typically operated in a slotted fashion. The length of the slot is such that a cell can be transmitted and propagated from an input port to an output port. At the beginning of a slot, all input ports that have a cell to send out start transmitting. At the end of the slot, the transmitted cells have all been switched to their output ports, assuming that no blocking occurred.

To alleviate the blocking problem, several solutions have been proposed. The simplest solution is to block the transmission of a cell that cannot proceed through a crosspoint. This occurs when a cell, while moving vertically towards its output port, arrives at a crosspoint which has already been set to the bar state by another cell. In this case, a blocking signal is returned on a reverse path to its input port. The input port stops transmitting, and it retries in the next slot. An alternative solution is to provide an arbiter who decides when each input port transmits so that there is no contention. The arbiter can be centralized, or each output port can have its own arbiter. In both solutions, it is necessary that each input port is equipped with a buffer, where cells can wait until they are successfully transmitted out. This makes the cross-bar into an input buffering switch.

The main problem with an input buffering switch is the lost throughput due to a situation that may occur referred to as the *head-of-line blocking*. This happens when a cell at the top of an input buffer is blocked. The cells behind it cannot proceed to their destination output port, even if these destination output ports are free. We demonstrate this using the 4×4 cross-bar shown in Figure 6.5. The number shown in each position of an input buffer indicates the destination output port of the cell held in that position. We can see that the destination output port of the cell at the top of each input buffer is the output port number 1. In this case, only one of the four cells will be transmitted. The remaining cells will be blocked, which means that the cells behind them will also be blocked even if they are not going to the same output port. It also means that only one path through the fabric will be used, while the other three will remain unused.

Simulation studies have shown that the maximum throughput of an input buffering cross-bar switch is 0.586 per slot per output port. This result was obtained assuming that

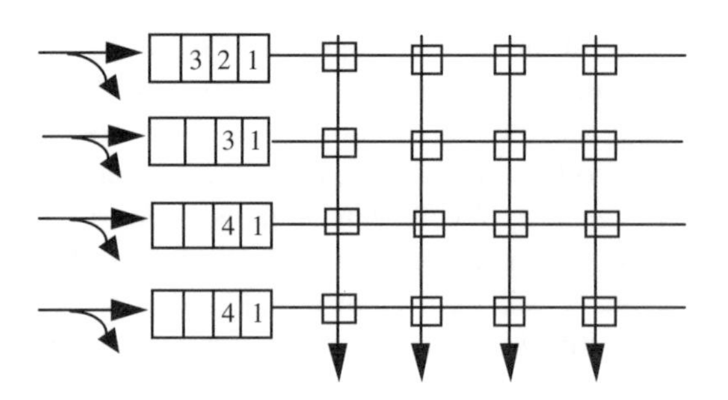

Figure 6.5 A 4 × 4 cross-bar with input buffering.

the input buffers always have at least one cell to transmit, and that the probability a cell chooses a destination output port is $1/N$, where N is the number of ports.

6.2.2 Banyan Networks

A Banyan network is constructed by interconnecting 2 × 2 switching elements. The switching elements are arranged in rows and columns, known as *stages*. Stages are numbered in an ascending order from left to right. Rows are also numbered in an ascending order from the top down. An example of a Banyan network is shown in Figure 6.6. Banyan networks are modular and they can be implemented in VLSI.

A Banyan network is an example of a multi-stage interconnection network. These networks may take different forms, depending on how the switching elements are interconnected, and on the number of input and output ports of the switching element. In Section 6.2.3, we will examine another multi-stage interconnection network, known as the *Clos network*.

In a Banyan network, there exists only a single path between an input port and an output port. Up to N paths can be established between input and output ports, where N is the number of ports. The output ports of the Banyan network are addressed in binary

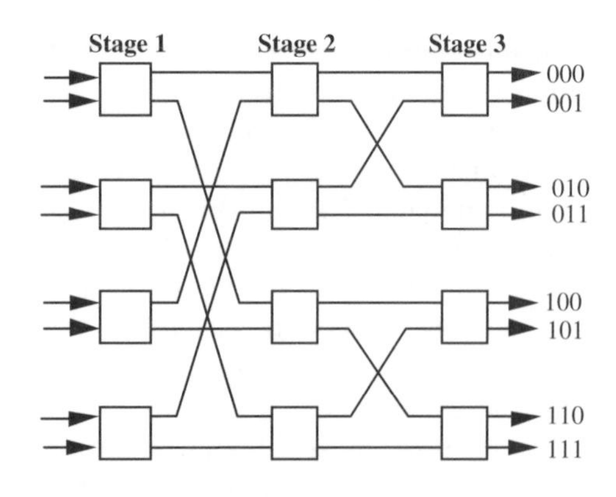

Figure 6.6 An 8 × 8 Banyan network.

in an ascending order. For instance, in the 8×8 Banyan network shown in Figure 6.6, there are eight output ports and they are numbered in binary as follows: 000, 001, 010, 011, 100, 101, 110, and 111.

A 2×2 switching element consists of two inlets, referred to as the upper and the lower inlet, and two outlets, referred to as the upper and the lower outlet, and it is *bufferless*, i.e. it does not have an internal buffer for holding cells. When two cells are offered to a switching element, one through the upper inlet and the other through the lower inlet, both cells can be switched if one is going to the upper outlet and the other to the lower outlet. A collision will occur if both incoming cells seek the same outlet, either the upper or the lower one. A Banyan network consisting of bufferless switching elements is known as a *Bufferless Banyan network.*

Each cell is self-routed through the Banyan network as follows. Let us consider a cell in an input port, which is about to be launched into the Banyan network. The binary address of its destination output port is attached in front of the cell in the reverse order. At each switching element, a cell coming in from the upper or lower inlet is routed to the upper outlet if its leading bit is 0 and to the lower outlet if it is 1. The leading bit is then dropped. For example, let us assume that a cell in the first input port is about to be launched into the Banyan network shown in Figure 6.6. Its destination is output port 011. The bit string 110 is attached in front of the cell, and it is used for the self-routing function. At the switching element in the first stage, the cell is routed to the upper outlet, and the leading bit is dropped. The attached bits are now 11. At the switching element in the second stage, the cell is routed to the lower outlet and the leading bit is dropped. At this moment, only one attached bit is left which is 1. At the last switching element in stage three, the cell is routed to the lower outlet and the leading bit is dropped. The cell has now established a path from its input port to the destination output port. The cell holds the path until the entire cell is transmitted and propagated through the Banyan network. Typically, only a few bits of the cell occupy each link along the established path.

In this Banyan network, two different paths may have to share a common link. For instance, in Figure 6.6 the path from the first input port to output port 001 and the path from the fifth input port to output port 000 share the same link between the second and third stages. This is an example of internal blocking. External blocking can also occur when two different paths have the same destination output port. Due to internal and external blocking, the throughput of this network is worse than that of a crossbar, assuming uniformly distributed destinations. In view of this, all switch architectures which are based on Banyan networks use various mechanisms for overcoming either internal blocking or external blocking or both. Such mechanisms are: input buffering with cell deflection, buffered switching elements, multiple copies of a Banyan network, and removing internal and external blocking through a sorter. Below, we examine each of these solutions.

Input buffering with cell deflection

An example of this architecture is shown in Figure 6.7. Each input port is equipped with a buffer, which can accommodate a finite number of cells. Cells are stored in the input buffer upon arrival from the incoming link. A cell is lost if it arrives at a time when the input buffer is full. Each input port is managed by a cell processor, indicated in Figure 6.7 with a circle. The cell processor carries out the header conversion and attaches the self-routing bits. It is also responsible for transmitting the cell into the fabric. The Banyan

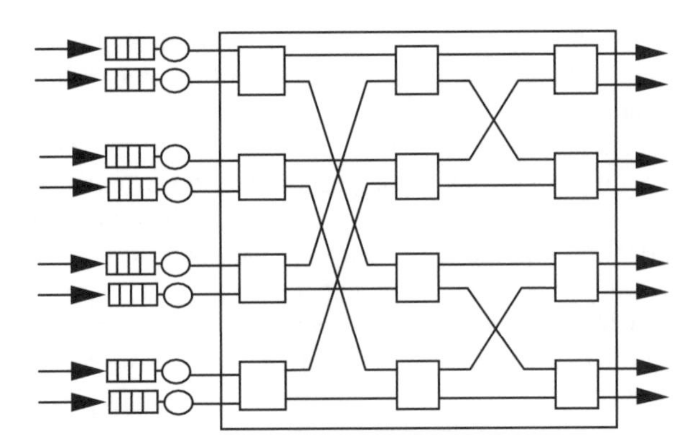

Figure 6.7 Input buffering Banyan switch with cell deflection.

network is bufferless and it is operated in a slotted fashion. A slot is long enough so that a cell can be completely transmitted and propagated to its destination output port. At the beginning of each slot, each cell processor launches a cell into the fabric. If two cells collide onto the same output port of a switching element, then one of the two cells will go through and the other will be blocked. This is done by sending back to the appropriate cell processor a blocking signal. The cell processor stops transmitting the cell, and it waits until the next slot to launch its cell again. In the worst case, a cell processor may have to try for several slots before it succeeds to launch a cell through the switch fabric successfully. This scheme has similar problems as the cross-bar with input buffering, i.e. its throughput is reduced to head-of-line blocking.

Buffered switching elements

In this architecture, each switching element is equipped with an internal buffer for holding entire cells. Dedicated buffers can be placed at the input ports, or at the output ports, or at both input and output ports of the switching element. Also, instead of dedicated buffers, a memory can be used which can be shared by all input and output ports of the switching element. A Banyan network consisting of buffered switching elements is known as a *buffered Banyan network.*

Below, we discuss the operation of a buffered Banyan switch consisting of buffered switching elements with input and output buffers. The north-west corner of the switch is shown in Figure 6.8. Each input port of the switching element has a buffer that can accommodate one cell, and each output port has a buffer that can accommodate m cells, where m is typically greater than one. The switch is operated in a slotted manner, and the duration of the slot is long enough so that a cell can be completely transferred from one buffer to another. All transfers are synchronized, so they all start at the beginning of a slot and they all end at the end of the slot. Also, all the buffers function so that a full buffer will accept another cell during a slot, if the cell at the head of the buffer departs during the same slot.

A cell is switched through the fabric in a store-and-forward fashion. Upon arrival at the switch, a cell is delayed until the beginning of the next slot. At that time it is forwarded to the corresponding input buffer of the switching element in the first stage, if the buffer is empty. The cell is lost if the buffer is full. This is the only place where cell loss can occur.

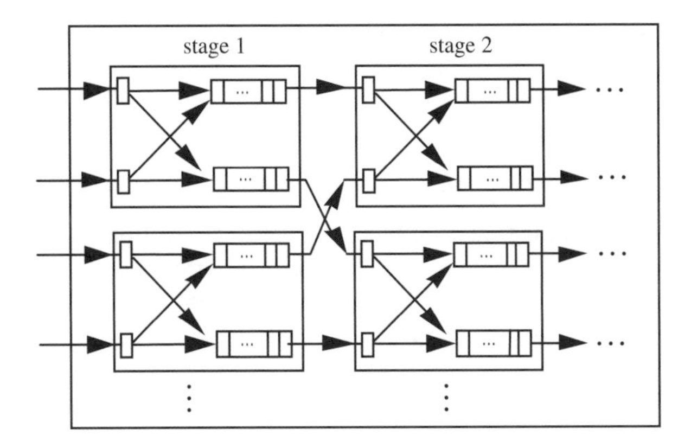

Figure 6.8 The NW corner of a buffered Banyan network.

A cell in an input buffer of a switching element is forwarded to its destination output buffer of the same switching element if there is a free space. If there is no free space, the cell is forced to wait in its input buffer until the end of the slot. At the beginning of the next slot, it will again attempt to move to its destination output buffer. If both input buffers contain a cell destined for the same output buffer in the switching element, and this output buffer has only one empty space, then one cell chosen randomly from the two input buffers will be forwarded. The transfer of a cell at the head of the output buffer of a switching element to the input buffer of the switching element in the next stage is controlled by a backpressure mechanism. If the input buffer of the next switching element is free, then the cell is forwarded to the switching element. Otherwise, the cell waits until the next slot and it tries again. Due to the backpressure mechanism, no cell loss can occur within the buffered Banyan switch.

The buffered Banyan switch alleviates internal and external blocking. However, it can only accommodate a fixed number of cells in its switching elements. If an output port becomes *hot* (it receives a lot of traffic), then a bottleneck will start building up in the switching element associated with this output port. Cell loss can eventually occur, as the bottleneck will extend backwards to the input ports of the switch.

To minimize queueing at the beginning stages of the buffered Banyan switch, another buffered Banyan switch is added in front of it, referred to as the *distribution network*, as shown in Figure 6.9. The distribution network is an identical copy of the buffered Banyan switch, which is now referred to as the *routing network*, with the exception that cells in each switching element are routed alternatively to each of the two outputs. This

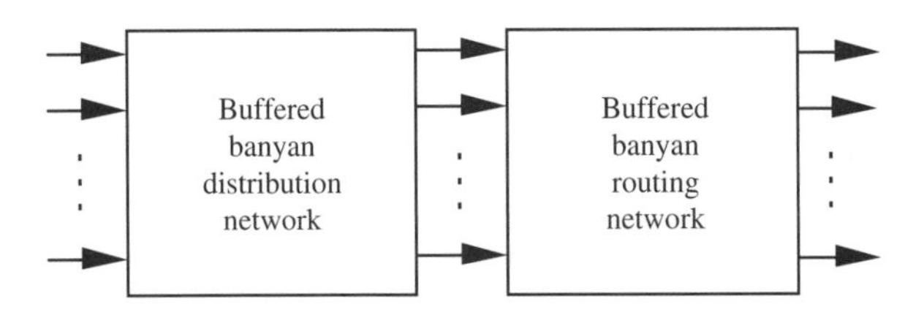

Figure 6.9 A buffered Banyan switch with a distribution network.

mechanism distributes the traffic offered to the routing network randomly over its input ports, which has the effect of minimizing queueing in the switching elements at the beginning stages of the routing network.

Multiple copies of a Banyan network

An alternative way to minimize internal blocking is to use multiple identical copies, say K copies, of a bufferless Banyan network, as shown in Figure 6.10. Typically, K is smaller than the total number of input/output ports of the switch. This scheme can be seen as an extension of the bufferless Banyan switch with input buffering and cell deflection, described above.

A packet processor can launch a cell on any of the K Banyan networks. The packet processor is busy during the initial time when the leading bits of a cell are propagated through the fabric. Once the cell has established its path to the output port, the packet processor becomes free, and it can launch another cell on a different Banyan network. Deflection is still possible within each Banyan network, and therefore a small queue at each input port is necessary. When a cell is deflected on a Banyan network, the packet processor will attempt to launch it on a different Banyan network. A packet processor can launch up to K cells, one per Banyan network. These K cells may be all directed to the same output port or to different output ports, depending upon their destination. Also, a maximum of K cells may be in the process of arriving at any of the output ports, and therefore, output port buffers are required.

The Batcher-Banyan switch

This type of switch consists of a *Batcher sorter* linked to a Banyan network via a shuffle-exchange, as shown in Figure 6.11. The Batcher sorter is a multi-stage interconnection network that orders the cells offered to it according to their destination output port address. That is, if N cells are offered to the Batcher sorter, one per input port, then these N cells will appear at the output ports of the Batcher sorter in an ascending order, as follows. The cell with the lowest destination output port address will appear at the top output port, then the cell with the next lowest destination output port address will appear at the

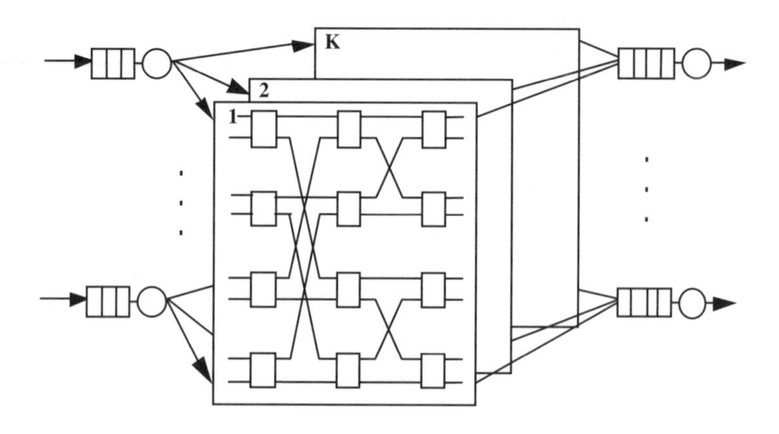

Figure 6.10 Multiple copies of Banyan networks.

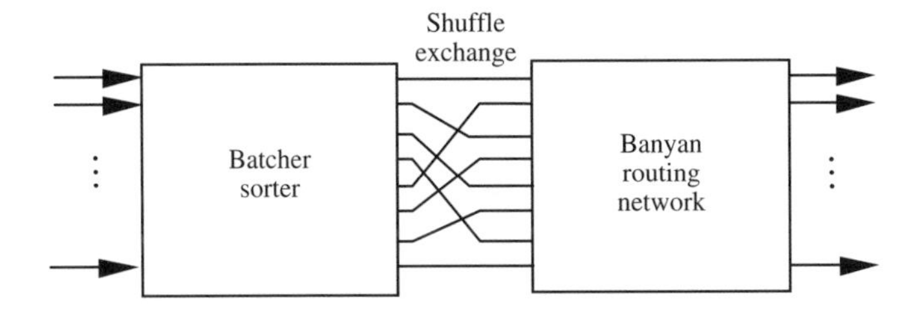

Figure 6.11 A Batcher-Banyan switch.

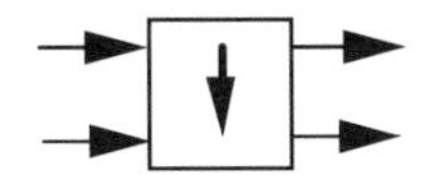

Figure 6.12 A Batcher switching element.

second output port, and so on. Two cells with the same destination output port address will appear at adjacent output ports.

The Batcher sorter consists of 2×2 switching elements, as shown in Figure 6.12. Each switching element compares the addresses of the two incoming cells, and it switches the one with the largest address to the output port indicated by the arrow. The other cell is switched to the other output port. If both cells have the same address, then the incoming cell on the upper (lower) input port is switched to the upper (lower) output port. If only a single cell is present, then it is switched as if it had the lowest address. That is, it is switched to the output port not indicated by the arrow.

An example of a Batcher sorter is shown in Figure 6.13. Three cells with switch destination output ports 8, 3 and 1 are presented to the network. (The switch destination output ports refer to the output ports of the Banyan network linked to the Batcher sorter.) Using the above routing rules, we see that these three cells will eventually appear at the three top output ports of the Batcher sorter sorted in an ascending order according to their destination output ports. It has been shown that the Batcher sorter produces *realizable*

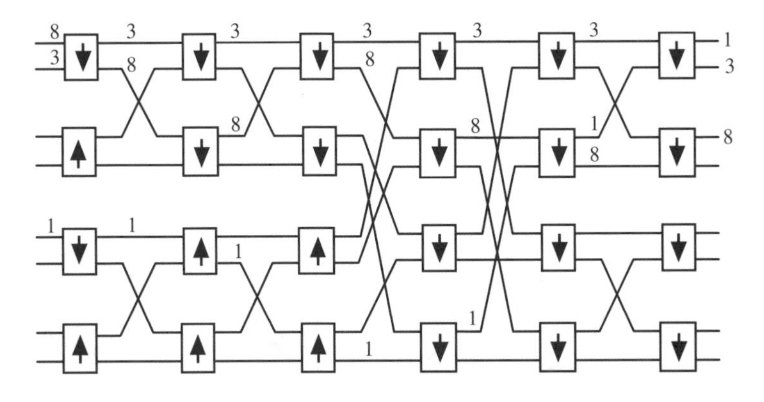

Figure 6.13 A Batcher sorter.

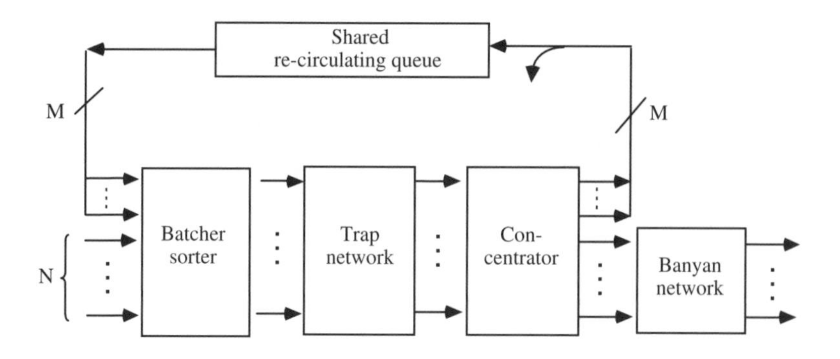

Figure 6.14 The starlite switch.

patterns. That is, the order in which the cells appear at the output ports of the Batcher sorter forms a pattern that can be switched by the Banyan network without any internal blocking. This is true if the cells presented to the Batcher network have different switch output port destinations.

The *starlite* switch, shown in Figure 6.14, is based on the Batcher-Banyan scheme, but it has an additional mechanism for eliminating external conflicts. This mechanism makes use of the fact that cells with the same switch destination output port appear at consecutive output ports of the Batcher sorter. Consequently, one of them can be forwarded to the Banyan network while the remaining cells are trapped. This is done in the trap network. All the trapped cells are concentrated in the top M lines, and the selected cells are concentrated in the remaining N lines which are connected to the Banyan routing network. The trapped cells are re-circulated into the switch fabric. Buffering is provided for recirculating cells. Recirculated cells have priority over new cells so as to maintain the order in which cell were transmitted. Cell loss can take place within the re-circulating queue. Also, cell loss can take place at any of the N incoming links of the Batcher sorter.

A variation of the starlite switch is the *sunshine* switch shown in Figure 6.15. This switch consists of K multiple Banyan networks instead of a single network employed in

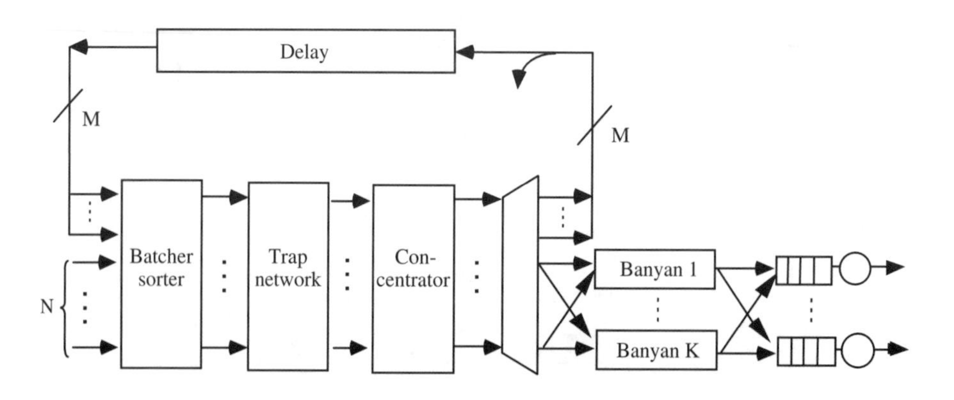

Figure 6.15 The sunshine switch.

the starlite switch. In view of this, up to K cells with the same switch output address can be selected by the network and forwarded to the Banyan networks. Due to the multiple Banyan networks, a buffer is required at each output port with a size of at least K cells. This switch has a lower recirculating rate than the starlite switch.

6.2.3 Clos Networks

A Clos network is another example of a multi-stage interconnection network. Unlike Banyan networks, Clos networks allow multiple paths between an input port and an output port. A three-stage Clos network is constructed as follows. The first stage consist of k $n \times m$ switching elements, i.e. each switching element has n input ports and m output ports, where $m > n$. The second stage consists of m $k \times k$ switching elements, and the third stage consists of k $m \times n$ switching elements. The switching elements are interconnected as follows. The output port i of the jth switching element in stage s, is connected to the jth input port of the ith switching element in the $(s + 1)$st stage, $s = 1, 2$. This Clos network provides m different paths between an input port of a switching element in the first stage and an output port of a switching element in the third stage.

An example of a three-stage Clos network, consisting of 6 2×4 switching elements in the first stage, 4 6×6 switching elements in the second stage, and 6 4×2 switching elements in the third stage, is shown in Figure 6.16. We observe that the number of ports of a switching element in the second stage is equal to the number of switching elements in the first stage, and the number of switching elements in the second stage is equal to the number of output ports of a switching element in the first stage. Also, the third stage is symmetric to the first stage. Finally, we note that for each input port of a switching element in the first stage, there are four different paths to the same output port of a switching element in the third stage, and the number of paths is equal to the number of output ports of a switching element in the first stage.

When a new connection is being set-up through a Clos network switch, a routing algorithm is used to calculate a path through its switch fabric. This is necessary, since

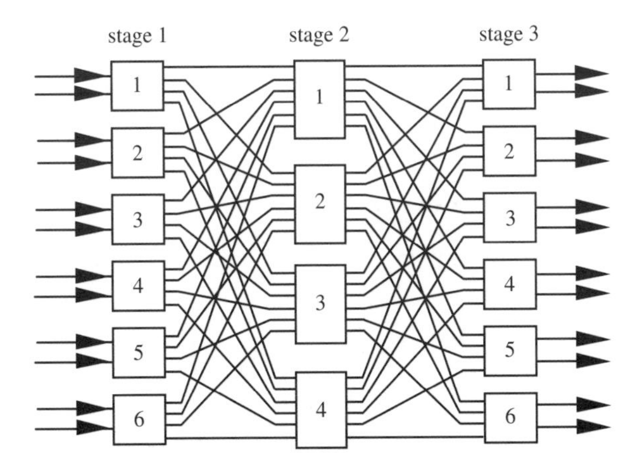

Figure 6.16 A three-stage Clos network.

there are multiple paths between an input and an output port. All the cells belonging to this connection follow the same path when traversing the Clos network.

6.2.4 Switch Architectures with N^2 Disjoint Paths

So far, we have examined different schemes designed to alleviate the problem of internal and external blocking. An alternative switch architecture which does not suffer from internal or external blocking can be obtained by allowing each input port of the switch to have its own dedicated link to each output port of the switch. Therefore, the total number of independent paths that can be set-up between the input and output ports is N^2, where N is the number of ports. The most-well known architecture of this class is the *knock-out* switch. This was an experimental switch, and it was quite complicated. A considerably simpler switch architecture of the same class is the *cross-bar tree* architecture shown in Figure 6.17. As can be seen, it consists of N planes, each interconnecting an input port of the switch to all the N output ports using 1×2 switching elements. The switch is operated in a slotted manner, and each cell is self-routed to its destination. Output buffers are placed at each output port, since more than one cell may arrive in the same slot.

6.3 SHARED MEMORY ATM SWITCH ARCHITECTURES

This is a very popular ATM switch architecture and its main feature is a shared memory that is used to store all the cells coming in from the input ports. The cells in the shared memory are organized into linked lists, one per output port, as shown in Figure 6.18. The shared memory is dual-ported, i.e. it can read and write at the same time. Currently, memories can handle up to 5 Gbps. At the beginning of a slot, all input ports that have a cell write it into the shared memory. At the same time, all output ports with a cell to transmit read the cell from the top of their linked list and transmit it out. If N is the number of input/output ports, then in one slot, up to N cells can be written into the shared memory and up to N cells can be transmitted out of the shared memory. If the speed of transmission on each incoming and outgoing link is V, then the switch can keep up at maximum arrival rate, if the memory's bandwidth is at least $2NV$.

The total number of cells that can be stored in the memory is bounded by the memory's capacity B, expressed in cells. Modern shared memory switches have a large shared

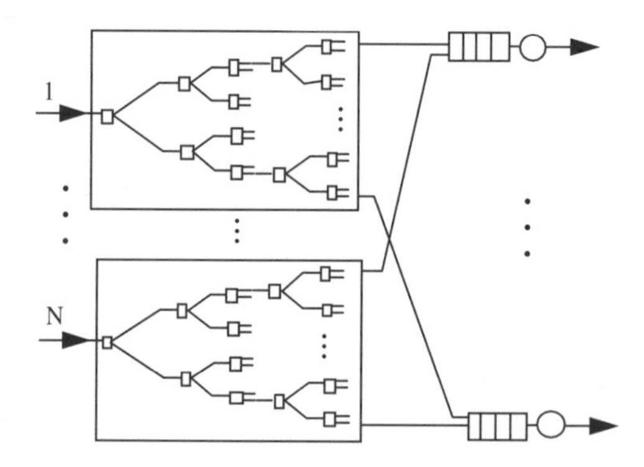

Figure 6.17 A cross-bar tree architecture.

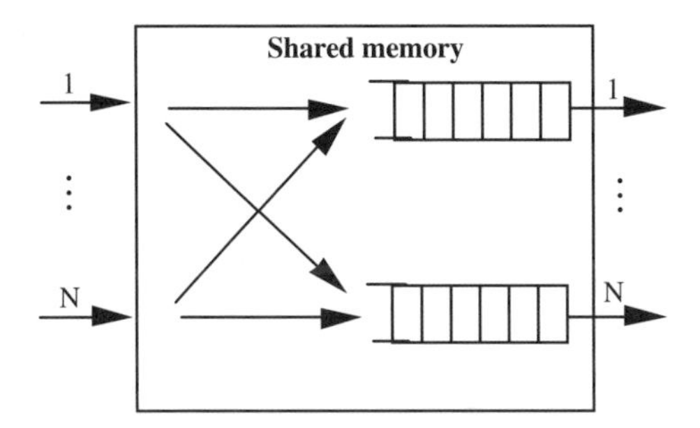

Figure 6.18 A shared memory switch.

memory and they can hold hundred of thousands of cells. The total number of cells allowed to queue for each output port i is limited to B_i, where $B_i < B$. That is, the linked list associated with output port i cannot exceed B_i. This constraint is necessary in order to avoid starvation of other output ports when output port i gets *hot*. An output port gets hot when a lot of the incoming traffic goes to that particular port. When this happens, it is possible that the linked list associated with the hot output port may grow to the point that it takes over most of the shared memory. In this case, there will be little space left for cells destined to other output ports. Typically, the sum of the B_i capacities of all linked lists is greater than B. More complicated constraints can also be used. For instance, each linked list i may be associated with a minimum capacity LB_i in addition to its maximum capacity B_i, where $LB_i < B$. LB_i is a dedicated buffer for output port i, and it is never shared with the other output ports. The sum of the *LBi* capacities of all linked lists is less than B.

Cell loss occurs when a cell arrives at a time when the shared memory is full, that is, it contains B cells. Cell loss also occurs when a cell with destination output port i arrives at a time when the total number of cells queued for this output port is B_i cells. In this case, the cell is lost, even if the total number of cells in the shared memory is less than B.

A large switch can be constructing by interconnecting several shared memory switches. That is, the shared memory switch described above is used as a switching element, and all the switching elements are organized into a multistage interconnection network.

An example of a shared memory switch is that shown in Figure 6.19, and it was proposed by Hitachi. First, cells are converted from serial to parallel (S/P), and header conversion (HD CNV) takes place. Subsequently, cells from all input ports are multiplexed and written into the shared memory. For each linked list, there is a pair of address registers (one to write, WA, and one to read, RA). The WA register for linked list i contains the address of the last cell of list i, which is always empty. The incoming cell is written in that address. At the same time, an address of a new empty buffer is read from the IABF chip, which keeps a pool of empty buffer locations, to update WA. Similarly, at each slot a packet from each linked list is identified through the content of the RA register, retrieved, demultiplexed and transmitted. The empty buffer is returned to the pool, and RA is updated with the next cell address of the linked list. Priorities may be implemented, by maintaining multiple linked lists, one for each priority, for the same output port.

One of the problems associated with the construction of early shared memory switches was that memories did not have the necessary bandwidth to support many input ports

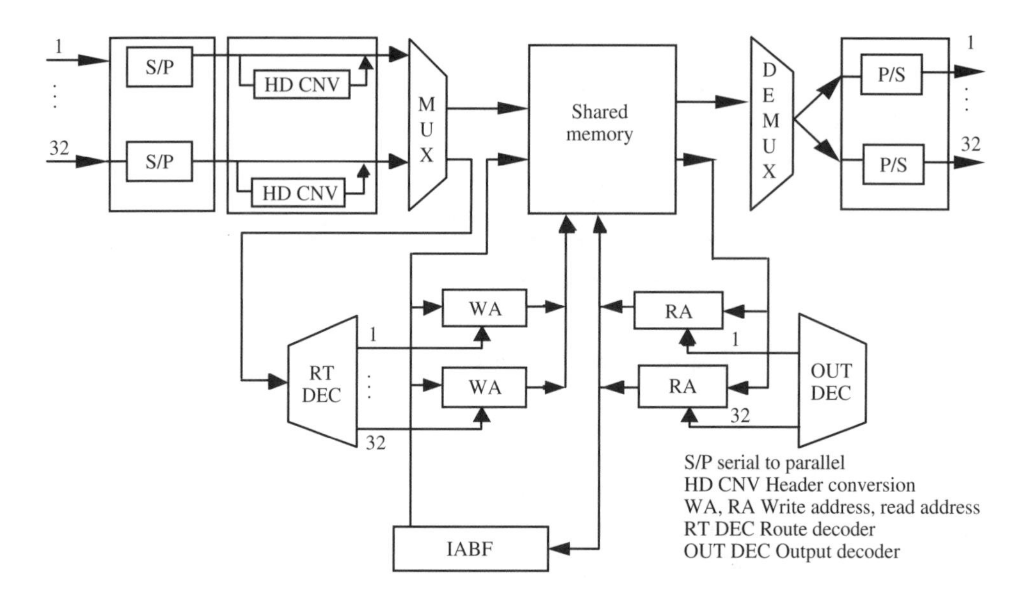

Figure 6.19 The Hitachi shared memory switch.

at a high speed. As a way of getting round this technical problem, a scheme known as bit-slicing was used in the Hitachi switch, in which K shared memories were employed instead of a single one. An arriving cell is divided into K subcells, and each subcell is written into a different shared memory at the same time in the same location. As a result, a cell in a linked list is stored in K fragments over the K shared memories. All the pointers for the linked list are the same in the K memories. Transmitting a cell out of the switch requires reading these K fragments from the K memories. Now if we assume that we have N links, and that the speed of each incoming and each outgoing link is V, then the bandwidth that each shared memory is required to have is $2NV/K$, rather than $2NV$, as in the case of a single shared memory.

The shared memory switch architecture has also been used in a nonblocking switch with output buffering, as shown in Figure 6.20. Instead of using a dedicated buffer for each output port, a shared memory switch is used to serve a number of output ports. The advantage of this scheme is the following. When using a dedicated buffer for each output port, free space in the buffer of one output port cannot be used to store cells of another output port. This may result in poor utilization of the buffer space. This problem is alleviated with multiple output ports sharing the same memory.

6.4 SHARED MEDIUM ATM SWITCH ARCHITECTURES

In this architecture, the input ports are connected to the output ports via a high-speed parallel bus, as shown in Figure 6.21. Each output port, indicated in Figure 6.21 by a circle, is connected to the bus via an interface and an output buffer. The interface is capable of receiving all cells transmitted on the bus, and through an Address Filter (A/F) it can determine whether each cell transmitted on the bus is destined for its output port. If a cell is destined for its output port, the cell is written in the output buffer, from where

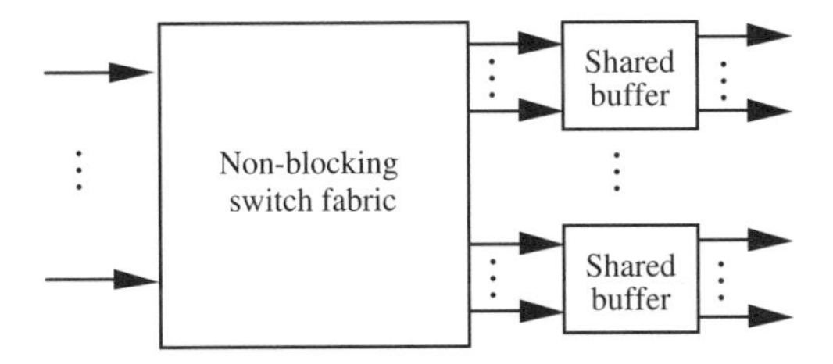

Figure 6.20 Shared memory used for a group of output ports.

it is eventually transmitted out by the output port. Cell loss can occur when a cell arrives at an output buffer at a moment when the buffer is full.

It is possible that the input ports may be equipped with small buffers where the incoming cells can wait until they are successfully forwarded to their destination output port. Back-pressure may be employed between the output and the input buffers, to avoid losing cells at the output buffers. In this case, cell loss can only occur at the input ports of the switch, when a cell arrives at an input port at a time when its input buffer is full.

The bus is slotted and it has a bandwidth equal to NV, where N is the number of ports and V is in the speed of an incoming or an outgoing link. There are as many bus slots as the number of input ports, and the order in which the input ports are served by the bus is determined by a scheduler. The simplest scheduler is based on time-division multiplexing.

The N bus slots are organized into a frame, and the frame repeats for ever. Each input port owns a specific bus slot within the frame, and it can transmit one cell per frame when its bus slot comes up. If the input port has no cell to transmit, its bus slot goes by unused. A slightly more sophisticated scheduler that avoids the problem of unused bus slots is the *modified time-division multiplexing algorithm*. In this algorithm, the bus slots are not organized into a frame, and each input port does not own a bus slot within the frame. The scheduler only serves the input ports that have a cell to transmit in a cyclic manner. Input ports that do not have a cell to transmit are skipped. The algorithm works

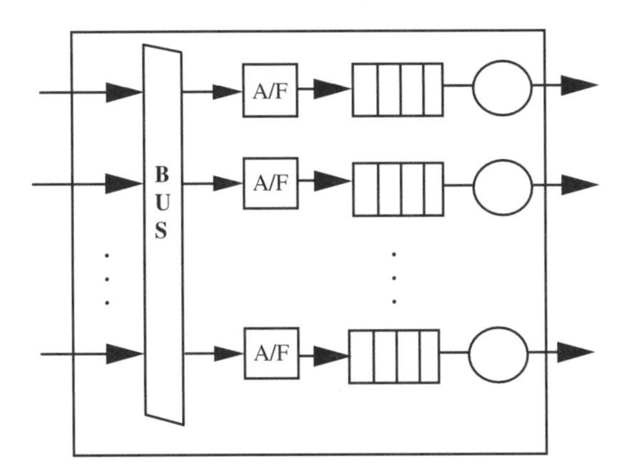

Figure 6.21 A shared medium switch architecture.

as follows. Assuming that it has just served input port i, it attempts to serve the next input port $i+1$ (mod N). If this input port has a cell to transmit, then the cell is transmitted in the next bus slot. If the input port has no cell to transmit, then the bus slot is not wasted. The scheduler searches the input ports after input port $i+1$ (mod N) sequentially, and allocates the bus slot to the first input port that it will find that has a cell to transmit. Other more complex algorithms can also be defined.

An example of a shared medium switch is the ATOM switch shown in Figure 6.22. To achieve the required speed on the shared bus, bit-sliced organization is employed. An incoming cell is converted into a bit-serial stream, which is then divided into P parallel streams, each feeding one of the P parallel subswitches. Routing of cells to output queues is performed in an identical manner over the P subswitches. This is achieved using a centralized address controller, which processes the headers of the incoming cells and instructs the subswitches as to which output buffer each cell is destined for. Specifically, the headers are extracted and routed to the address controller. The address controller multiplexes the headers and broadcasts them to the address filters of each output port. Each address filter determines which of the cells are to be written into its respective input buffer, and send the appropriate write control signals to the subswitches.

Large ATOM switches can be constructed by interconnecting shared medium switching elements so as to form a multi-stage interconnection network.

6.5 NONBLOCKING SWITCHES WITH OUTPUT BUFFERING

As mentioned earlier on in this chapter, an ATM switch whose input ports are equipped with buffers, irrespective of whether its output ports have buffers or not, is referred to as an input buffering switch. It is referred to as an output buffering switch if only its output ports have buffers. In Section 6.2.1, we discussed how a cross-bar switch with input buffers suffers from head-of-line blocking. This type of blocking occurs in all switches with input buffering, and it may cause the time it takes for a cell to traverse the switch to increase, which has the effect of decreasing the throughput of the switch. Output buffering

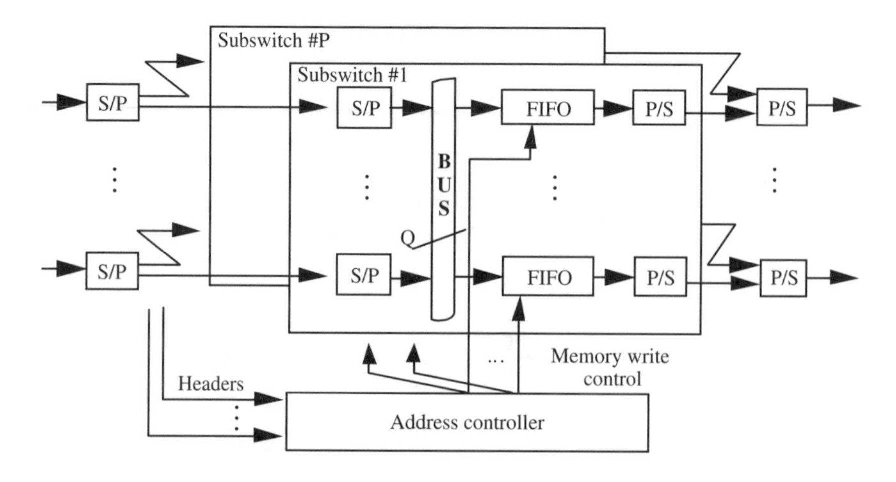

Figure 6.22 The ATOM switch architecture.

switches do not suffer from head-of-line blocking, and in view of this, they are preferable to input buffering switches.

ATM switches are also classified to *blocking* and *nonblocking* switches. In a nonblocking switch, its switching fabric does not cause internal or external blocking. On the other hand, in a blocking switch, its switching fabric may cause internal or external blocking, and therefore cells may collide while they are traversing the switch. In general, nonblocking switches are preferable to blocking switches.

A good example of a nonblocking switch with output buffering is the switch architecture with N^2 disjoint paths, discussed in Section 6.2.4. In this switch, each input port has its own dedicated path to each output port, thus eliminating the possibility of internal and external blocking. Also, there are no buffers at the input ports, which eliminates head-of-line blocking.

The shared memory switch, discussed in Section 6.3, is also a nonblocking switch with output buffering. In this case, the output buffers are the linked lists formed in the shared memory.

Finally, the shared medium ATM switch architecture without input buffers, discussed in Section 6.4, is also a nonblocking ATM switch with output buffering.

6.6 MULTICASTING IN AN ATM SWITCH

ATM connections may be point-to-point or point-to-multipoint. A point-to-point connection is used by two end devices to communicate over an ATM network, and a point-to-multipoint connection is used for multicasting and broadcasting. Multicasting is different to broadcasting. In a multicast, an end device, called the *root*, transmits the same information to a specific group of end devices, called the *leaves*. The root and the leaves are members of a particular multicast group, which is a subset of the set of all end devices attached to the network. In a broadcast, the root transmits the same information to all the end devices attached to the network.

An ATM switch should be able to carry both point-to-point and point-to-multipoint connections. In a point-to-point connection, an ATM switch simply transfers incoming cells belonging to the connection to a specific output port of the switch. In a point-to-multipoint connection, an ATM switch transfers incoming cells belonging to the connection to a number of different output ports of the switch. The number of leaves in the multicast may change during the time that the connection is up. New leaves may be added and existing ones may be dropped. These changes may result in changes in the set of destination output ports of an ATM switch to which the incoming cells should be delivered.

Various solutions have been proposed to introduce multicasting into the switch architectures examined in the previous sections. In the case of multi-stage interconnection switch architectures, two different approaches can be used to implement multicasting, namely the *time* and *space* approaches. In the time approach, each packet processor transmits each multicast cell to all the destination output ports of the switch, one at a time. That is, if a multicast cell should be distributed to four different output ports, then the packet processor will transmit it four times, each time to a different target output port. This is an acceptable solution when the size of the multicast group is small, but it introduces large delays when the multicast group is large. In the space approach, the cell is duplicated using a copy network. For instance, let us consider the buffered Banyan switch shown in Figure 6.9. It consists of a distribution network and the routing network. This switch can be seen as a point-to-point switch, i.e. it can deliver incoming cells belonging to a point-to-point connection to a specific output port of the switch. Multicasting can be introduced using a copy network in front of the distribution network. The structure of the copy network is the same as that of the distribution and routing networks, i.e. it consists of 2×2 buffered switching elements which are interconnected in a Banyan network. Each switching element in the copy network can generate up to two copies of each passing cell. The copy network performs cell duplication according to the size of the multicast group. Each duplicated cell goes through the distribution network, and then it is switched to its destination output port by the routing network. The copy network can modify the number of copies and their destination output port as leaves are added or dropped.

The shared memory switch architecture is more amenable to supporting multicasting than a multi-stage interconnection switch architecture. Various schemes for transmitting a multicast cell out of a number of output ports have been proposed. The simplest scheme is to copy a multicast cell to all the linked lists associated with the destination output ports of its multicast group. This method may result in higher memory usage, particularly when the multicast group is very large. Another method is to keep a separate linked list for multicast cells. In an $N \times N$ shared memory switch, the N linked lists, one per output port, are used to hold cells belonging only to point-to-point connections. In addition to these N linked lists, a new linked list is introduced which holds all the multicast cells. This mulitcast linked list is not served at the same time as the N linked lists. Also, during the time that cells are transmitted out of the multicast linked list, no cells are transmitted out of the N point-to-point linked lists. Transmission of a multicast cell is organized in such a way that it is transmitted out of all its destination output ports simultaneously. Various policies can be used to decide how the switch distributes its time between serving the N linked lists and serving the multicast linked list.

The shared medium switch architecture is probably the most suitable architecture for transmitting multicasting traffic. This is because a cell transmitted over the bus can be received by all output ports. The address filter of an output port will accept a multicast cell if it recognizes its address. The problem associated with this switch is that an output buffer associated with a multicast may become full. If backpressure is used, the transmitter will not be able to complete its multicast to all the target output ports. This problem can be resolved by keeping a list of all the output ports that did not receive the multicast, and keep multicasting the cell only to the output ports in the list.

6.7 SCHEDULING ALGORITHMS

Early ATM switches were equipped with very small buffers. For instance, in an output buffering switch, each output buffer may have had a capacity of less than 100 cells. The cells were served (i.e. transmitted out of the switch) in the order in which they came, that is, in a First-In-First-Out (FIFO) fashion. The FIFO scheduling algorithm does not take into account priorities among the cells, and the fact that cells may belong to different connections with different quality-of-service parameters. Therefore, a cell belonging to a connection used for a file transfer will be served first if it arrives before a cell belonging to a connection which is carrying data from a real-time application, such as voice or video. The justification for the FIFO algorithm was based on the fact that the queueing delay in a buffer was small, since the buffer size was small. Therefore, there was no need for a complex scheduling algorithm that would serve cells from different connections with different priorities. Another consideration in support of the FIFO algorithm was that it was easy to implement.

It became apparent, however, that it is possible to implement effectively complex scheduling algorithms that can manage large buffers with many queues, so that different connections could be served according to their requested quality of service. A number of different schedulers have been proposed and implemented. Below, we discuss some of these scheduling algorithms.

Static Priorities

Let us consider a nonblocking switch with output buffering, as shown in Figure 6.23. Each output buffer holds cells that belong to different connections that pass through this buffer. Each of these connections is associated with a quality-of-service category signaled to the switch at call set-up time. The cells belonging to these connections can be grouped into queues, one per quality-of-service category, and these queues can be associated with different scheduling priorities.

As will be seen in the following chapter, several quality-of-service categories have been defined. Let us consider the following four categories: *Constant Bit Rate* (CBR), *Real-Time Variable Bit Rate* (RT-VBR), *Non-Real-Time Variable Bit Rate* (NRT-VBR) and *Unspecified Bit Rate* (UBR). The CBR service category is intended for real-time applications which transmit at a constant bit rate, such as unencoded video and circuit emulation. The RT-VBR service category is intended for real-time applications which transmit at a variable bit rate, such as encoded voice and video. The NRT-VBR service category is for applications that transmit at variable bit rate, and do not have real-time requirements. The UBR service category is intended for delay-tolerant applications that do not require any guarantees, such as data transfers.

Using these four quality-of-service categories, each output buffer can be organized into four different queues, as shown in Figure 6.24. An arriving cell at an output buffer joins one of the four queues according to the quality-of-service category of the connection to which it belongs.

These queues can be assigned static priorities, which dictate the order in which they are served. These priorities are called static because they do not change over time, and they are not affected by the occupancy levels of the queues. For instance, the CBR queue has the highest priority, the RT-VBR the second highest priority, and so on, with the UBR queue having the lowest priority. These queues can be served as follows. Upon

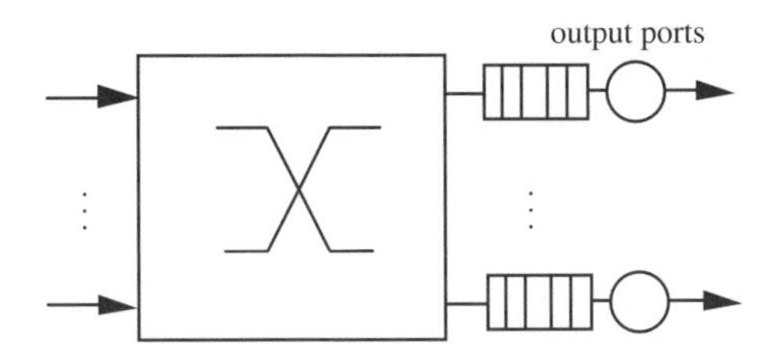

Figure 6.23 A nonblocking switch with output buffering.

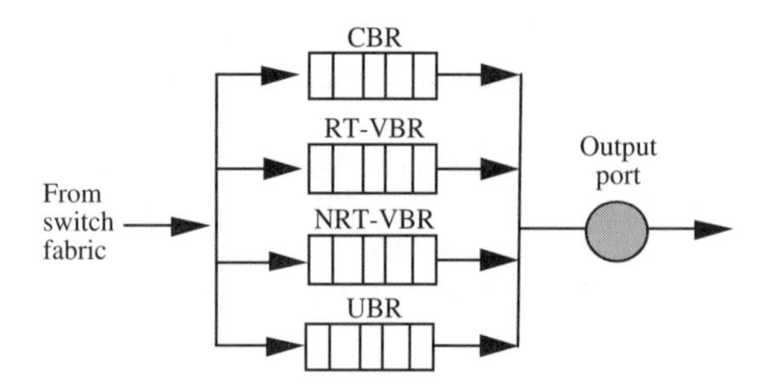

Figure 6.24 Logical queues for an output port.

completion of the transmission of a cell, the next cell for transmission is selected from the CBR queue. If the CBR queue is empty, the next cell for transmission is selected from the RT-VBR queue. If this queue is empty, then the next cell is selected from the NRT-VBR queue, and so on. If all queues are empty, then no cell will be selected for transmission. Thus, in essence, the CBR queue is served until it becomes empty. Then, the next priority queue is served until it becomes empty, and so on. If, during the time that a cell from, say, the UBR queue is being transmitted out, a cell arrives in one of the higher-priority queues (i.e. the CBR queue, RT-VBR queue or NRT-VBR queue), then this high priority cell will be transmitted out next after the transmission of the cell from the UBR queue is completed.

Additional scheduling rules can be introduced which take into account the current status of the queues. A typical example of such a rule is the 'aging factor'. If a queue, typically a low priority queue, has not been served for a period of time which is longer that a prespecified threshold, then the queue's priority is momentarily raised so that some of its cells can be transmitted out.

Early Deadline First (EDF) Algorithm

In this algorithm, each cell is assigned a deadline upon arrival at the buffer. This deadline indicates the time by which the cell should depart from the buffer. It is calculated by adding a fixed delay to the arrival time of the cell. This delay may vary according to the quality-of-service category of the cell. The scheduler serves the cells according to their deadlines, so that the one with the earliest deadline gets served first. A cell that is assigned a deadline closer to its arrival time will suffer a low delay in the buffer. On the other hand, a cell that is assigned a deadline far away from the time that it arrived at the buffer may suffer a longer delay before it gets transmitted out.

Using this scheme, cells belonging to delay-sensitive applications, such as voice or video, can be served first by assigning them deadlines closer to their arrival times.

The Weighted Round-Robin Scheduler

Each output buffer is organized into a number of queues. For instance, there could be one queue for each connection that passes through the particular output port. There could also be fewer queues, such as one queue per quality-of-service category. The scheduler serves one cell from each queue in a round robin fashion. The queues are numbered from 1 to M, and they are served sequentially. That is, if a cell from queue 1 was just served, then the next queue to serve is queue 2. This sequential servicing of the queues continues until the Mth queue is served, whereupon it goes back to queue 1. If the next queue to be served, say queue i, is empty the scheduler skips it and goes on to queue $i + 1$. (This algorithm is similar to the modified time-division multiplexing algorithm described in Section 6.4.)

Weighted round-robin scheduling can be used to serve a different number of cells from each queue. For instance, let us assume that there are five connections with weights: 0.1, 0.2, 0.4, 0.7 and 1. Multiplying each weight by a common number so that all the weights become integer, in this case by 10, gives: 1, 2, 4, 7 and 10. The scheduler will serve one cell from the first queue, two from the second queue, four from the third queue, seven from the fourth queue, and ten from the fifth queue.

Now, let us consider the case where one of the queues, say queue 5, becomes idle for a period of time. Then, queues 1–4 will be served as before, and queue 5 will be skipped each time its turn comes up. The ten slots that would have been used for queue 5 are now used for queues 1–4 proportionally to their weights.

6.8 THE LUCENT AC120 SWITCH

In this section, we describe the basic architecture of an existing product, the Lucent AC120 switch. The switch was designed to operate at the edge of an ATM network, and it is equipped with interfaces for Ethernet, DS-1, DS-3, E1, E2 and OC-3. The switch combines features from both the shared memory switch architecture and the medium shared switch architecture. As shown in Figure 6.25, the switch consists of a number of I/O cards which are attached to two buses. Each bus runs at 600 Mbps. There is also a CPU attached to the bus, which is used for call management. Each I/O card can receive cells from both buses, but it can only transmit on one bus. Half of the I/O cards transmit onto the same bus, and the other half onto the second bus. If one bus develops an error, all the I/O cards switch to the other bus. The buses are slotted and each bus slot carries an ATM cell with an added proprietary header of three bytes. Transmission on each bus takes place using the modified time-division multiplexing algorithm described in Section 6.4. That is, transmission is done in a round-robin fashion among those I/O cards that have a cell to transmit. Each I/O card transmits for one time slot.

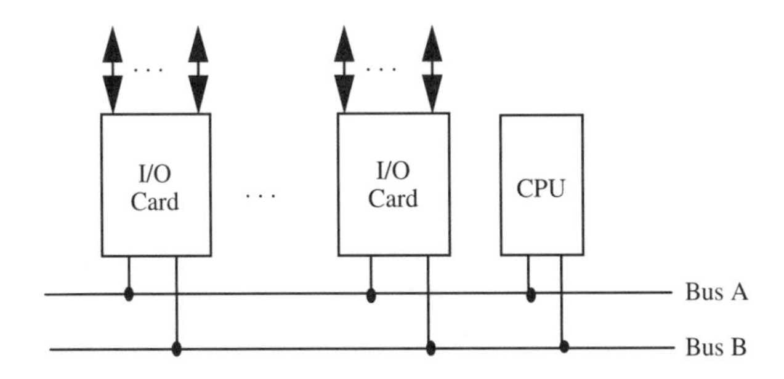

Figure 6.25 The architecture of the LDR200.

Some of the relevant components of an I/O card are shown in Figure 6.26. Each I/O card has I/O devices, a DRAM main memory and a CPU which controls all memory read/write functions, cell queueing, and management of the I/O card. An I/O card can receive cells from both buses, and under normal conditions it transmits only to one bus. It can also receive and transmit cells from its dual input/output ports, which are the actual input/output ports of the switch. The FIFO queues associated with the two buses are used to receive/transmit cells from/to the bus at the rate of 600 Mbps. Therefore, the total rate at which cells may arrive at both input FIFOs (assuming that no cells are being transmitted out) can be as high as 1.2 Gbps.

Cells are transferred from/to the input FIFOs to/from the shared memory using a Direct Memory Access (DMA) scheme. It is possible that an input FIFO may become full. Backpressure is then used to protect against cell loss. This is done by instructing the other I/O cards not to transmit to the card experiencing congestion.

An input and output FIFO also serve all the duplex ports of the I/O card. Cells are written from/to the input FIFO to/from the shared memory at a rate matching the speed of the I/O ports. The switch supports the following interfaces: 1 OC-3, 2 DS-3, 6 DS-1, 4 E1, 6 E2 or 5 Ethernet ports. Of the five Ethernet ports, four are 10 Mbps ports, and the fifth one can be configured either as 100 Mbps or as a 10 Mbps port.

We now briefly examine the set of queues maintained in the shared memory, which has a configurable capacity of up to one million cells. There are three queues for all the input ports, and 10 queues per output port, as shown in Figure 6.27. An incoming cell from any of the input ports is queued into one of three queues, namely CBR, VBR-1 and VBR-2. These three queues are served to completion on a priority basis, with the CBR queue having the highest priority and the VBR-2 queue having the lowest. Traffic coming in from the two buses is queued into one of the 10 queues of the destination output port. Four of these queues are used for CBR traffic, namely, CBR-1, CBR-2, CBR-3 and CBR-4. The next five queues are used for VBR traffic, namely VBR-1, VBR-2, VBR-3, VBR-4 and VBR-5. Finally, the last queue is used for UBR traffic. A proprietary scheduling algorithm, known as *AqueMan*, is used to schedule the order in which the cells are transmitted out of these 10 queues. The algorithm utilizes static priorities with

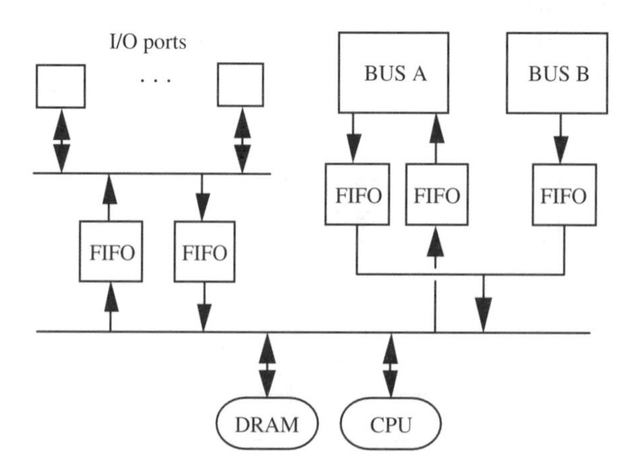

Figure 6.26 The architecture of an I/O card.

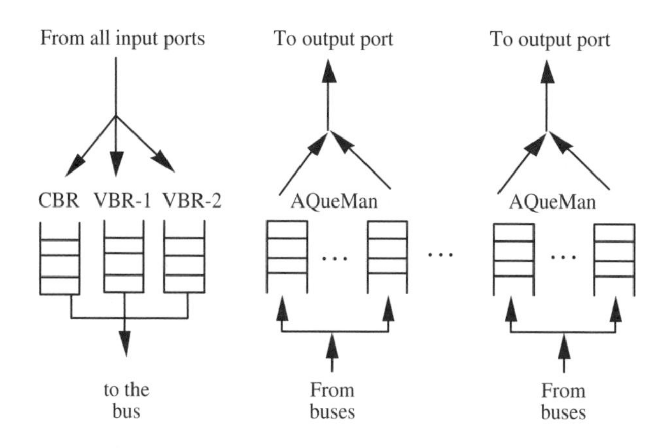

Figure 6.27 Queueing in the shared memory.

additional scheduling rules based on the status of the queues. Specifically, queues CBR-1, CBR-2, CBR-3, CBR-4 and VBR-1 have top priority, i.e. they are served to completion before the other five queues can be served. CBR-1 has the highest priority, CBR-2 has the next highest priority, and so on, with VBR-1 having the lowest priority among these five queues. When these five queues are empty, the scheduler will transmit a cell from the remaining five queues (i.e. VBR-2, VBR-3, VBR-4, VBR-5 and UBR), using an algorithm that takes into account the time a queue was served last (aging factor) and the number of cells in a queue (queue depth factor).

An I/O card is considered congested when the number of cells in its shared memory reaches a high water mark. At this point, entire queues are purged until the number of cells drops below a low water mark.

6.9 PERFORMANCE EVALUATION OF AN ATM SWITCH

The evaluation of the performance of an ATM switch is an important step prior to building a new switch or deploying a switch in the field. Typically, one is interested in quantifying the cell loss probability as a function of the load offered to the input ports. Other performance measures such as jitter and delay are also be of interest. It is relatively difficult to quantify jitter and delay, and in view of this, most of the performance studies of switch architectures have focused on calculating the cell loss probability.

A performance evaluation of a switch can be done by experimentation or by using modeling techniques. Experimentation involves carrying out real traffic measurements on a switch, whereas modeling involves building a simulation or a queueing model, which is then manipulated to obtain the desired performance measures. To carry out traffic measurements on a switch, one should able to reproduce realistic traffic loads, and also the switch has to *exist*. Modeling is used when the switch does not exist, or when the specific configuration that will be employed in the field is not available.

Simulation techniques are easy to apply, and one does not need to have good knowledge of queueing theory, though some knowledge of queueing theory may help design better simulation models! Typically, a switch is dimensioned so that it will have a very low cell loss probability. This probability could be as low as 10^{-8}, i.e. on average there is one lost cell for every 100 000 000 arriving cells. One needs extremely long simulation runs to obtain a good statistical estimate of such a low probability. For instance, if no cell loss is observed after simulating a switch for 100 000 000 arriving cells, then that does not mean that the average cell loss probability is zero! One needs to simulate the switch for a number of arriving cells which is considerably larger than 100 000 000 to obtain a reliable statistical estimate. In view of this, estimating very low cell loss probabilities by simulation is not practical, since it may be extremely time-consuming. Specialized techniques, such as *rare event simulation*, can be used to reduce the length of the required simulation run.

Queueing-based models, on the other hand, can run very quickly on a workstation, and they can be used to estimate very low cell loss probabilities. However, they are more difficult to develop, and they require good knowledge of queueing theory.

The way that one depicts the cell arrival pattern in a model, affects the performance of the switch. In many simulation and queueing-based models of ATM switch architectures, it was assumed that the arrival of cells at an input port was Poisson or Bernoulli distributed. Also, a uniform distribution of cells over the output ports was assumed. That is, if there are N output ports, then an arriving cell seeks each of these N output ports with probability $1/N$. These assumptions make a switch architecture look good, i.e. it can carry traffic loads at high link utilization without loosing too many cells. However, these arrival patterns are unrealistic, and they should not be used in performance evaluation studies, since they give optimistic results. In general, ATM traffic tends to be bursty and correlated (see Section 7.1), and it should not be modeled by a Poisson or a Bernoulli arrival process. As far as the output port destination probabilities are concerned, it makes sense to assume that a cell from input port i will go to output port j with some probability p_{ij}. This permits us to test different destination patterns and cases where one or more output ports become *hot*. Also, one can assume that an entire train of cells goes to the same output port.

PROBLEMS

1. Draw an 8×8 Banyan network, number the output ports, and give an example of how a cell can be self-routed from an input port to an output port.
2. Why are output buffering switches preferable to switches with input buffering?
3. In the buffered Banyan switch described in Section 6.2.2, is it possible to loose cells inside the switch fabric? Why?
4. Does the addition of a buffered Banyan distribution network in front of a buffered Banyan network eliminates external blocking? Why?
5. Consider the Batcher sorter network shown in Figure 6.13. The following eight cells with destinations 1, 4, 5, 3, 2, 1, 5, 8 are offered to the sorter at the same time. The first cell with destination 1 is offered to the top input port of the sorter, the second cell with destination 4 is offered to the second input port of the sorter, and so on. In what order they will appear at the output ports of the sorter?
6. Consider the switch architecture shown in Figure 6.17.
   (a) Explain why it is nonblocking.
   (b) Why is it necessary to have buffers at the output port?
7. Consider a shared memory switch architecture, and assume that the total number of cells that can be accommodated in the switch is B. Why is there an upper bound, less than B, on the size of each linked list?

8. Consider an $N \times N$ shared memory switch. The speed of each incoming link and each outgoing link is V Mbps. An incoming cell is first processed (i.e. header conversion, addition of a tag in front of the cell used for internal purposes), then it is written in the memory. All incoming cells from the input ports are processed by the same CPU. How much time the CPU has to process a cell in the worst case, i.e. when cells arrive back-to-back on all input ports? (Assume that the CPU is not busy when a read or write to the shared memory is performed.)
9. Consider an $N \times N$ shared memory switch. The speed of each incoming link and each outgoing link is V Mbps. What is the minimum speed of the memory in order to switch all the incoming traffic?
10. Consider an $N \times N$ shared medium switch. The speed of each incoming link and each outgoing link is V Mbps. What should be the speed of the bus in order to keep up with its incoming links?

APPENDIX: A SIMULATION MODEL OF AN ATM MULTIPLEXER–PART 1

The objective of this project is to simulate the queueing model shown in Figure 6.A1 that represents an output buffer of a nonblocking 4×4 ATM switch. This queueing model is known as the *ATM multiplexer*. For simplicity, we will assume that all the arrival processes are Bernoulli. In a follow-up simulation project described in the next chapter, you will develop this simulation model further by introducing more realistic traffic models.

You can either write your own simulation program following the instructions given below, or use a simulation language.

Project Description

The buffer has a finite capacity of B cells, and it receives cells from four different arrival streams, numbered 1 to 4. Each arrival stream is associated with an input port, and represents the stream of traffic from that input port destined for the output port under study. The buffer is served by a server which represents the output port.

Each arrival stream is slotted, i.e. time is divided into equal time periods, known as *slots*. A slot is long enough so that a cell can completely arrive at the queue. Also, the time it takes to serve a cell from a queue by the output port (i.e. transmitted out) is also equal to one slot. We will assume that each arrival stream is a Bernoulli process, i.e. a slot carries a cell with probability p or it is empty with probability $1 - p$.

The slots of each arrival stream and the service slot are all synchronized, so by the end of a slot, up to four new cells (one per arrival stream) may arrive at the buffer and one cell may depart from the buffer. An arriving cell is first buffered and then it is transmitted

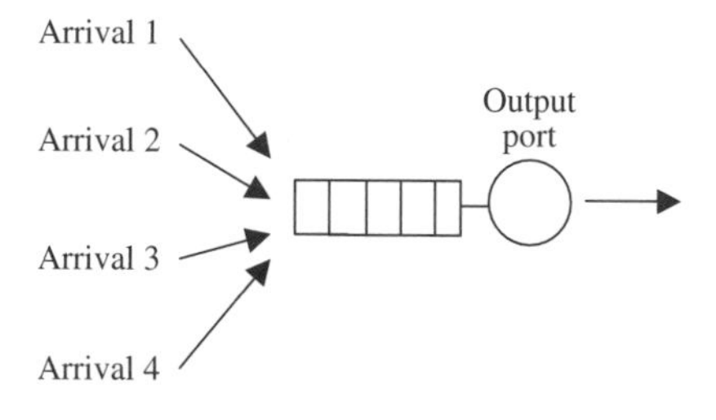

Figure 6.A1 The ATM multiplexer.

out. If a cell arrives at slot i to find the buffer empty, it will not depart from the buffer in the same slot. Rather, it will depart in slot $i + 1$.

Task 1

Write a simulation model to simulate the above system, with a view to estimating the cell loss probability as a function of p. A cell is lost if it arrives at a time when the buffer is full.

Plot several curves, each corresponding to a different value of B, to show how the cell loss probability varies as a function of p.

Task 2

Augment your simulation to allow for different Quality-of-Service (QoS) queues in the output buffer. Specifically, we will assume that an arriving cell joins one the following four queues: CBR queue, RT-VBR queue, NRT-VBR queue, and UBR queue. A cell joins the CBR queue with probability p_{CBR}, the RT-VBR with probability $p_{\text{RT-VBR}}$, the NRT-VBR queue with probability $p_{\text{NRT-VBR}}$, and the UBR queue with probability p_{UBR}.

The CBR queue has the highest priority and the UBR queue has the lowest. These queues are served by the output port using the static priority rule described in Section 6.7. That is, at the beginning of each slot, the cell to be transmitted out is selected from the CBR queue using the FIFO policy. If it is empty, then it is selected from the RT-VBR queue using the FIFO policy, and so on. If all queues are empty, then no cell is transmitted out.

The four queues share the total buffer space B associated with the output port. To make sure that each queue i does not grow too big so that it takes up most of the buffer space B, we impose a low bound L_i and an upper bound U_i. The upper bound is used to limit how big the ith queue can grow, and typically, it is selected so that the sum of the upper bounds of the four queues is larger than B. The lower bound L_i can be seen as a dedicated buffer permanently allocated to the ith queue. That is, the L_i buffer spaces are only used to store cells belonging to the ith queue. If L is the sum of the lower bounds of the four queues, then we have that $L < B$. Therefore, the total number of buffer spaces that can be shared by the four queues is $B - L$.

To clarify the use of the upper and lower bound, let us consider the ith queue, and let us assume that it is empty. The first cell that arrives at the ith queue, will be stored in the buffer (i.e. it will take up one of the L_i spaces), and the total number of dedicated buffers associated with the ith queue will be reduced by one. This will continue until all dedicated buffer spaces have been used up. When a cell arrives at the ith queue at a time when all dedicated buffer spaces have been used up, the cell will be accepted if the following two conditions are met:

(a) The total number of cells in the ith queue is less than its upper bound U_i.
(b) For each queue i, calculate the number of cells X_i which is over its lower bound L_i. (If the number of cells in the ith queue is less or equal to L_i, then $X_i = 0$.) Then, the sum $X_1 + X_2 + X_3 + X_4$ has to be less than $B - L$.

Calculate the cell loss probability for all the cells arriving at the ATM multiplexer, and for each QoS queue, as a function of p for given values of the probabilities p_{CBR}, $p_{\text{RT-VBR}}$, $p_{\text{NRT-VBR}}$ and p_{UBR}, and for given L_i and U_i, i = 1, 2, 3, 4.

Plot several curves, each corresponding to a different value of B, p_{CBR}, $p_{\text{RT-VBR}}$, $p_{\text{NRT-VBR}}$, p_{UBR} and L_i and $U_i, i = 1, 2, 3, 4$, to show how the cell loss probability varies as a function of p.

Structure of the Simulation Model

The simplest way to construct the simulation model is to use the unit-time design, with the unit time equal to a slot.

For task 1, do the following for each slot:

(a) If the buffer has at least one cell, then one cell departs. Update the total number of cells in the buffer.
(b) Draw a pseudo-random number r, where $0 < r < 1$. If $r < p$, then arrival stream 1 has a cell. Else, no arrival from stream 1.
(c) Repeat above step for arrival streams 2, 3 and 4.
(d) Calculate how many of these arriving cells will be admitted to the buffer, based on the current number of cells in the buffer. Keep a counter for the total number of arrivals and another counter for the total number of lost cells. Update the number of cells in the buffer.
(e) Simulate for a total number of arrived cells, N, and print out, the ratio: number of lost cells/N.

For Task 2, follow the same steps, but modify the rules for admitting a cell to the buffer and serving a cell from the buffer as explained in the previous section. Use the following procedure to determine which queue an admitted cell will join:

(a) Draw a pseudo-random number r, where $0 < r < 1$.
(b) If $r < p_{\text{CBR}}$, then the cell will join the CBR queue.
(c) If $p_{\text{CBR}} < r < p_{\text{CBR}} + p_{\text{RT-VBR}}$, then the cell will join the RT-VBR queue.
(d) If $p_{\text{CBR}} + p_{\text{RT-VBR}} < r < p_{\text{CBR}} + p_{\text{RT-VBR}} + p_{\text{NRT-VBR}}$, then the cell will join the VRT-VBR queue.
(e) If $r > p_{\text{CBR}} + p_{\text{RT-VBR}} + p_{\text{NRT-VBR}}$, then it will join the UBR queue.

(Remember to draw a new random number each time you need one!)

7

Congestion Control in ATM Networks

Congestion control, otherwise known in the ATM Forum standards as *traffic management*, is a very important component of ATM networks. It permits an ATM network operator to carry as much traffic as possible so that revenues can be maximized without affecting the quality of service offered to the users.

As we will see in this chapter, an ATM network can support several quality-of-service categories. A new connection at call set-up time signals to the network the type of quality-of-service category that it requires. If the new connection is accepted, the ATM network will provide the requested quality-of-service to this connection without affecting the quality-of-service of all the other existing connections. This is achieved using congestion control, in conjunction with a scheduling algorithm that is used to decide in what order cells are transmitted out of an ATM switch (see Section 6.7).

Two different classes of congestion control schemes have been developed: *preventive* and *reactive*. In preventive congestion control, as its name implies, we attempt to prevent congestion from occurring. This is done using the following two procedures: *Call (or connection) Admission Control* (CAC), and *bandwidth enforcement*. Call admission control is exercised at the connection level, and it is used to decide whether to accept or reject a new connection. Once a new connection has been accepted, bandwidth enforcement is exercised at the cell level to ensure that the source transmitting on this connection is within its negotiated traffic parameters.

Reactive congestion control is based on a totally different philosophy than preventive congestion control. In reactive congestion control, the network uses feedback messages to control the amount of traffic that an end device transmits so that congestion does not arise.

In this chapter, we first present the parameters used to characterize ATM traffic, the quality-of-service parameters, and the ATM quality-of-service categories. Then we describe in detail the preventive and the reactive congestion control schemes.

7.1 TRAFFIC CHARACTERIZATION

The traffic submitted by a source to an ATM network can be described by the following traffic parameters: *Peak Cell Rate* (PCR), *Sustained Cell Rate* (SCR), *Maximum Burst Size* (MBS), *burstiness* and *correlation of inter-arrival times*. Also, various probabilistic and empirical models have been used to describe the arrival process of cells. Below, we examine these traffic parameters in detail, and we also briefly introduce some empirical and probabilistic models. Two additional parameters, namely *Cell Delay Variation Tolerance* (CDVT) and *Burst Tolerance* (BT), will be introduced later on in this chapter.

Peak Cell Rate (PCR)

This is the maximum amount of traffic that can be submitted by a source to an ATM network, and it is expressed as ATM cells per second. Due to the fact that transmission speeds are expressed in bits per second, it is more convenient to talk about the peak bit rate of a source, i.e. the maximum number of bits per second submitted to an ATM connection, rather than its peak cell rate. The peak bit rate can be translated to the peak cell rate, and vice versa, if we know which ATM adaptation layer is used. The peak cell rate has been standardized by both the ITU-T and the ATM Forum.

Sustained Cell Rate (SCR)

Let us assume that an ATM connection is up for a period of time equal to D. During that time, the source associated with this connection transmits at a rate that varies over time. Let S be the total number of cells transmitted by the source during the period D. Then, the average cell rate of the source is S/D. (One would be inclined to use the abbreviation ACR for the Average Cell Rate, but this abbreviation is used to indicate the *Allowed Cell Rate* in the ABR mechanism described in Section 7.8.1!)

The average cell rate has not been standardized by ITU-T or by the ATM Forum. Instead, an upper bound of the average cell rate, known as the *Sustained Cell Rate* (SCR) has been standardized by the ATM Forum. This is obtained as follows.

Let us first calculate the average number of cells submitted by the source over successive short periods T. For instance, if the source transmits for a period D equal to 30 minutes and T is equal to 1 second, then there are 1800 T periods, and we will obtain 1800 averages, one per period. The largest of all these averages is called the sustained cell rate. We observe that the SCR of a source cannot be larger than the source's PCR, nor can it be less than the source's average cell rate.

The SCR is not to be confused with the average rate of cells submitted by a source. However, if we set T equal to D, then the SCR simply becomes the average cell rate at which the source submits cells to the ATM network. For instance, in the above example, the SCR will be equal to the average cell rate, if T is equal to 30 minutes. The value of T is not defined in the standards, but in the industry it is often taken to be equal to 1 second.

Maximum Burst Size (MBS)

Depending upon the type of the source, cells may be submitted to the ATM network in bursts. These bursts may be fixed or variable in size. For instance, in a file transfer, if the records retrieved from the disk are of fixed size, then each record results to a fixed number of ATM cells submitted to the network back-to-back. In an encoded video transfer, however, each coded image has a different size, which results to a variable number of cells submitted back-to-back. The Maximum Burst Size (MBS) is defined as the maximum number of cells that can be submitted by a source back-to-back at peak cell rate. The MBS was standardized by the ATM Forum.

Burstiness

This is a notion related to how the cells transmitted by a source are clumped together. Typically, a source is bursty if it transmits for a period of time and then becomes idle for another period of time, as shown in Figure 7.1. The longer the idle period, and the higher the arrival rate during the active period, the more bursty the source is.

The burstiness of a source can significantly affect the cell loss in an ATM switch. Let us consider an output buffer of the output buffering non-blocking ATM switch, shown in Figure 6.23. This buffer is shown in Figure 7.2. It has a finite capacity queue and it is served by a link indicated in the figure by a circle. The arrival stream of ATM cells to the queue can be seen as the superposition of several different arrival streams coming from the input ports of the switch. A cell that arrives at a time when the queue is full is lost.

Now, from queueing theory we know that as the arrival rate increases, the cell loss increases as well. What is interesting to observe is that a similar behavior can be also seen for the burstiness of a source. The curve in Figure 7.3 shows qualitatively how the cell loss rate increases as the burstiness increases while the arrival rate remains constant. (Detailed curves relating the cell loss probability to the burstiness of the arrival process can be obtained by carrying out the simulation project 'A simulation model of an ATM multiplexer—Part 2' described at the end of this chapter.)

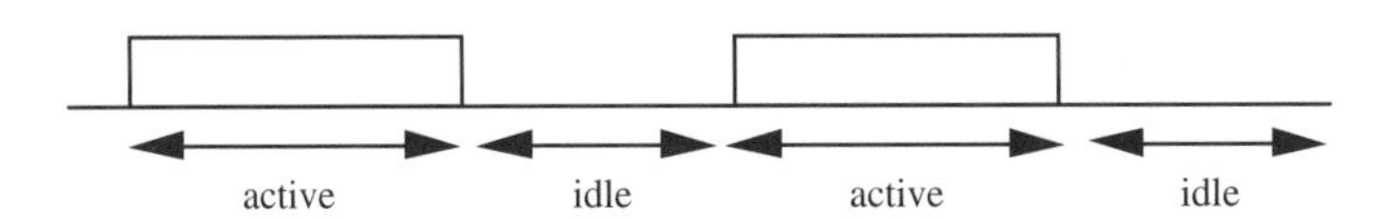

Figure 7.1 A bursty source.

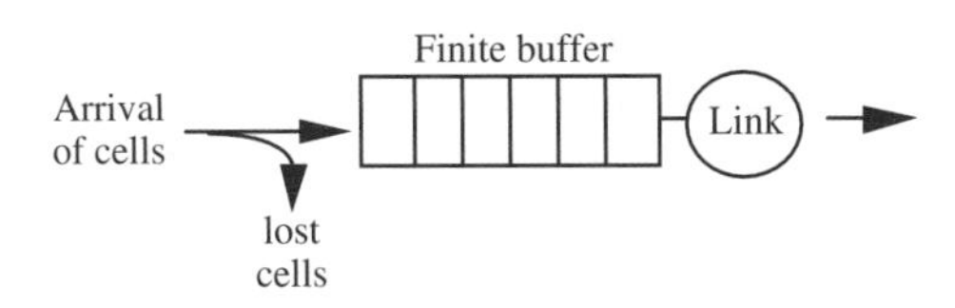

Figure 7.2 A finite capacity buffer.

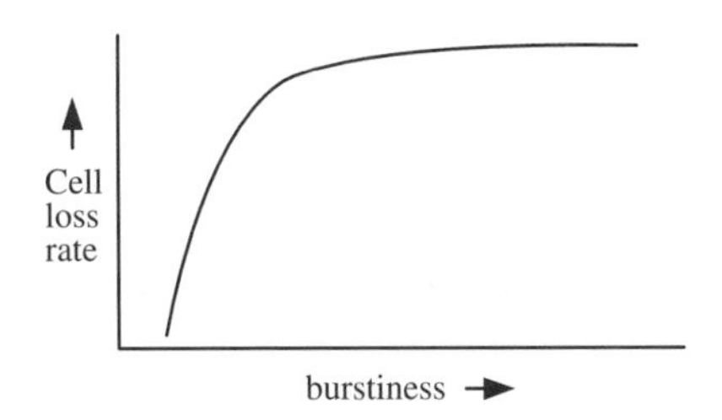

Figure 7.3 Cell loss rate vs. burstiness.

Correlation

Let us consider successive inter-arrival times of cells generated by a source, as shown in Figure 7.4. In an ATM environment, it is highly likely that the inter-arrival times are correlated either positively or negatively. Positive correlation means that if an inter-arrival time is large (or small), then it is highly likely that the next inter-arrival time will also be large (or small). Negative correlation implies the opposite. That is, if an inter-arrival time is large (or small), then it is highly likely that the next inter-arrival time will be small (or large). As in the case of burstiness, the correlation of the inter-arrival time of cells can significantly affect the cell loss probability in an ATM switch.

7.1.1 Standardized Traffic Descriptors

The ATM Forum has standardized the following traffic descriptors: peak cell rate, cell delay variation tolerance, sustained cell rate, and maximum burst size. The ITU-T has only standardized the peak cell rate. The peak cell rate, sustained cell rate, and maximum burst size depend upon the characteristics of the source. The cell delay variation tolerance is used in the *Generic Cell Rate Algorithm* (GCRA), discussed in Section 7.7.1, and it is independent of the characteristics of the source. It is specified by the administrator of the network to which the source is directly attached.

7.1.2 Empirical Models

Several empirical models have been developed to predict the amount of traffic generated by a variable bit rate MPEG video coding algorithm. These empirical models are statistical models, and they are based on regression techniques.

MPEG is a standards group in ISO that is concerned with the issue of compression and synchronization of video signals. In MPEG, successive video frames are compressed following a format like: I B B B P B B B P B B B I, where I stands for *I-frame*, B for *B-frame* and P for *P-frame*. An intra-coded frame, or I-frame, is an encoding of a picture based entirely on the information in that frame. A predictive-coded frame, or P-frame, is based on motion compensated prediction between that frame and the previous I- or P-frame. A bidirectional-coded frame, or B-frame, is based on motion compensated prediction between that frame and the previous I- or P-frame or the next I- or P-frame.

The encoder also can select the sequence of I, P and B frames, which form a group of frames known as a *Group Of Pictures* (GOP). The group of frames repeats for the entire duration of the video transmission.

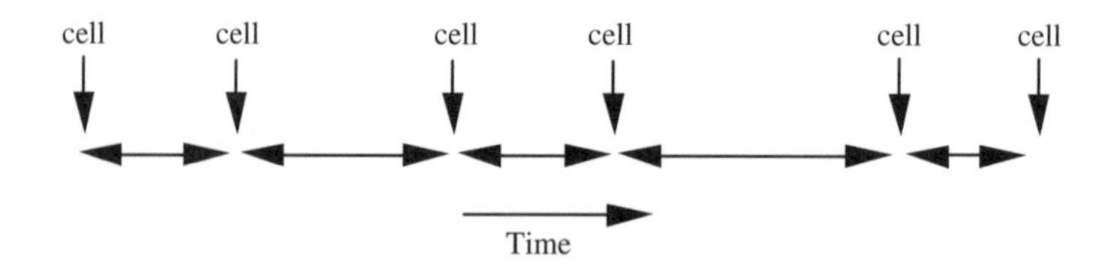

Figure 7.4 Successive inter-arrival times of cells.

The size of the resulting frame varies significantly between frame types. I-frames are the largest while B-frames are the smallest. The size of an I-frame varies based on picture content. P- and B-frames vary depending on the motion present in the scene, as well as picture content.

The number of bits produced by each frame in such a sequence is correlated, and it can be predicted using an *AutoRegressive-Moving Average* (ARMA) model. Such a model can be used in a performance evaluation study to generate video traffic. (See the simulation project 'Estimating the ATM traffic parameters of a video source', given at the end of this chapter.)

7.1.3 Probabilistic Models

Probabilistic models of arrival processes are abstractions of real-life arrival processes. They do not represent real-life arrival processes exactly, but they capture some of the traffic parameters described above, and in view of this, they are extremely useful in performance evaluation studies.

When we talk about a probabilistic model of an ATM arrival process, we assume that the arrival process is generated by a source which transmits cells over an ATM link. The link is assumed to be used exclusively by this source, and it is slotted with a slot being equal to the time it takes for the link to transmit a cell. Now, if we place ourselves in front of the link and observe the slots go by, then we will see that some of the slots carry a cell while others are empty. A model of an ATM arrival process describes which slots carry a cell and which slots are idle.

ATM sources are classified into *Constant Bit Rate* (CBR) and *Variable Bit Rate* (VBR). A CBR source generates the same number of bits every unit time, whereas a VBR source generates traffic at a rate that varies over time. Examples of CBR sources are circuit emulation services such as T1 and E1, unencoded voice and high quality audio. Examples of VBR sources are encoded video, encoded voice with suppressed silence periods, IP over ATM, and frame relay over ATM. The arrival process of a CBR source is easy to characterize. The arrival process of a VBR source is more difficult to characterize, and it has been the object of many studies.

CBR Sources

As mentioned above, a CBR source generates the same number of bits every unit time. For instance, a 64 Kbps unencoded voice produces 8 bits every 125 ms. Since the generated traffic stream is constant, the PCR, SCR and average cell rate of a CBR source are all the same, and a CBR source can be completely characterized by its PCR.

Let us assume that a CBR source has a PCR equal to 150 cells per second, and the ATM link over which it transmits has a speed, expressed in cells per second, of 300. Then, if we observe the ATM link, we will see that every other slot carries a cell. If the speed of the link is 450 cells per second, then every third slot carries a cell, and so on.

VBR Sources

A commonly used traffic model for data transfers is the *on/off process* shown in Figure 7.5. In this model, a source is assumed to transmit only during an active period, known as

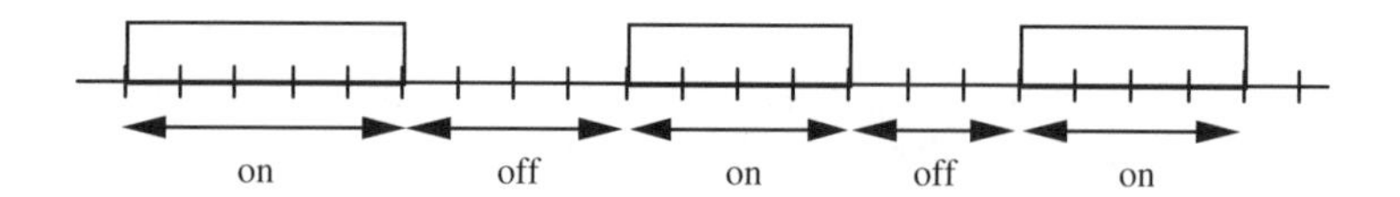

Figure 7.5 The on/off process.

the *on period*. This period is followed by a silent period, known as the *off period*, during which the source does not transmit. This cycle of an on period followed by an off period repeats continuously until the source terminates its connection. During the on period, there may be a cell transmitted every slot, or every fixed number of slots, depending upon the source's PCR and the speed of the link.

The PCR of an on/off source, is the rate at which it transmits cells during the on period. For example, if it transmits every other slot, then its PCR is equal to half the speed of the link, where the link's speed is expressed in cells per second. Alternatively, we can say that the source's peak bit rate is half the link's capacity, expressed in bits per second. The average cell rate is:

$$\frac{\text{PCR} \times \text{mean length of on period}}{\text{mean length of on and off period}}$$

The on/off model captures the notion of burstiness, which is an important traffic characteristic in ATM networks. The burstiness of a source is indicative of how cells are clumped together. There are several different ways of measuring burstiness. The simplest one is to express it as the ratio of the mean length of the on period divided by the sum of the mean on and off periods, that is

$$r = \frac{\text{mean on period}}{\text{sum of mean on and off periods}}$$

This quantity can be also seen as the fraction of time that the source is active transmitting. When r is close to 0 or to 1, the source is not bursty. The burstiness of the source increases as r approaches 0.5. Another commonly used measure of burstiness, but more complicated to calculate, is the squared coefficient of variation of the inter-arrival times defined by $\text{var}(X)/(E(X))^2$, where X is a random variable indicating the inter-arrival times.

The length of the on and off periods of the on/off process follow an arbitrary distribution. A special case of the on/off process is the well-known *Interrupted Bernoulli Process* (IBP), which has been used extensively in performance studies of ATM networks. In an IBP, the on and off periods are geometrically distributed and cells arrive during the on period in a Bernoulli fashion. That is, during the on period, each slot contains a cell with probability α or it is empty with probability $1 - \alpha$.

The IBP process can be generalized to the two-state *Markov Modulated Bernoulli Process* (MMBP). A two-state MMBP consists of two alternating periods, period 1 and 2. Each period is geometrically distributed. During period i, we have Bernoulli arrivals with rate α_i, $i = 1, 2$. That is, each slot during period i has α_i probability of containing

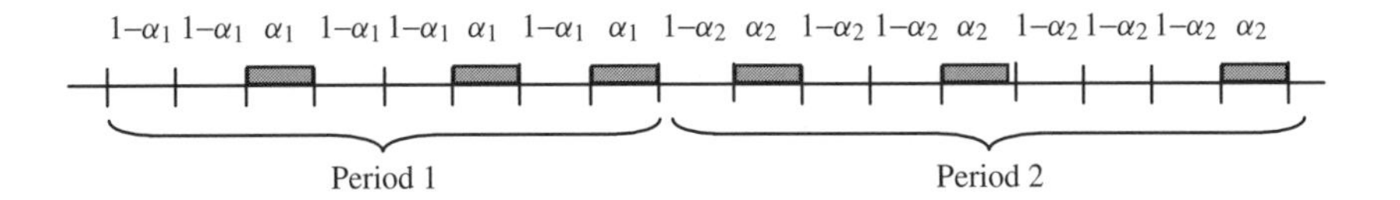

Figure 7.6 The two-state MMBP.

a cell, as shown in Figure 7.6. Transitions between the two periods are as follows:

	period 1	period 2
period 1	p	1-p
period 2	1-q	q

That is, if the process is in period 1 (period 2), then in the next slot it will be in the same period with probability $p(q)$ or it will change to period 2 (period 1)with probability $1 - p$ $(1 - q)$. A two-state MMBP model captures both the notion of burstiness and the correlation of inter-arrival times. More complicated MMBPs can be obtained using n different periods.

The above arrival processes were defined in discrete time. That is, we assumed that the link is slotted, and the length of the slot is equal to the time it takes to transmit a cell. Similar arrival processes have been defined in continuous time. In this case, the underlying assumption is that the link is not slotted, and the arrival of an ATM cell can occur at any time. The continuous-time equivalent of the IBP is the *Interrupted Poisson Process* (IPP), which is a well known process used in teletraffic studies. In an IPP, the on and off periods are exponentially distributed, and cells arrive in a Poisson fashion during the on period. An alternative model can be obtained using the fluid approach. In this case, the on and off periods are exponentially distributed as in the IPP model, but the arrivals occur during the on period at a continuous rate, like fluid flowing in. This model has been used extensively in performance studies, and it is referred to as the *Interrupted Fluid Process* (IFP).

The IPP can be generalized to a two-state *Markov Modulated Poisson Process* (MMPP), which consists of two alternating periods, period 1 and 2. Each period i, $i = 1, 2$, is exponentially distributed with a mean $1/\mu_i$ and during the ith period arrivals occur in a Poisson fashion at the rate of λ_i. More complicated MMPPs can be obtained using n different periods.

7.2 QUALITY OF SERVICE (Q*o*S) PARAMETERS

A number of different parameters can be used to express the quality of service of a connection, such as, *Cell Loss Rate* (CLR), *jitter*, *Cell Transfer Delay* (CTD), *peak-to-peak cell delay variation* and *Maximum Cell Transfer Delay* (max CTD).

The Cell Loss Rate (CLR) is a very popular quality-of-service parameter, and it was the first to be used in ATM networks. This is not surprising, since there is no flow control between two adjacent ATM switches or between an end device and the switch to which it is attached. Also, cell loss is easy to quantify, as opposed to other quality-of-service parameters such as jitter and cell transfer delay. Minimizing the cell loss rate in an ATM switch has been used as a guidance to dimensioning ATM switches, and also a large number of call admission control algorithms were developed based on the cell loss rate.

The jitter is an important quality-of-service parameter for real-time applications, such as voice and video. In these applications, the inter-arrival gap between successive cells at the destination end device cannot be greater than a certain value, as this may cause the receiving play-out process to pause. In general, the inter-departure gaps between successive cells transmitted by the sender are not the same as the inter-arrival gaps at the receiver. Let us consider Figure 7.7. The gap between the end of the transmission of the ith cell and the beginning of the transmission of the $(i + 1)$st cells is t_i. The gap between the end of the arrival of the ith cell and the beginning of the arrival of the $(i + 1)$st cell is s_i. The inter-departure gap t_i may be less than, equal to or greater than s_i. This is due to buffering and congestion delays in the ATM network. This variability of the inter-arrival times of cells at the destination is known as jitter.

It is important that the service provided by an ATM network for a voice or a video connection is such that the jitter is bounded. If the inter-arrival gaps s_i are less than the inter-departure gaps t_i, then the play-out process will not run out of cells. (If this persists for a long period of time, however, it may cause over-flow problems.) If the inter-arrival gaps are consistently greater than the inter-departure gaps, then the play-out process will run out of cells and will pause. This is not desirable, since the quality of the voice or video delivered to the user will be affected. Bounding jitter is not easy to accomplish.

The Cell Transfer Delay (CTD) is the time it takes to transfer a cell end-to-end, that is, from the UNI of the transmitting end device to the UNI of the receiving end device. It is made up of a fixed component and a variable component. The fixed cell transfer delay is the sum of all fixed delays that a cell encounters from the transmitting end device to the receiving end device, such as propagation delay, fixed delays induced by transmission systems, and fixed switch processing times. The variable cell transfer delay, known as the *peak-to-peak cell delay variation*, is the sum of all variable delays that a cell encounters from the transmitting end device to the receiving end device. These delays are primarily due to queueing delays in the switches along the cell's path. The peak-to-peak cell delay variation should not to be confused with the Cell Delay Variation Tolerance (CDVT), which is used in the Generic Cell Rate Algorithm (GCRA) described in Section 7.7.1.

The Maximum Cell Transfer Delay (max CTD) is another quality-of-service parameter that defines an upper bound on the end-to-end cell transfer delay. This upper bound is not an absolute bound, rather it is a statistical upper bound, which means that the actual end-to-end cell transfer delay may occasionally exceed max CTD. That is, the sum of the fixed cell transfer delay and the peak-to-peak cell delay variation may exceed max CTD,

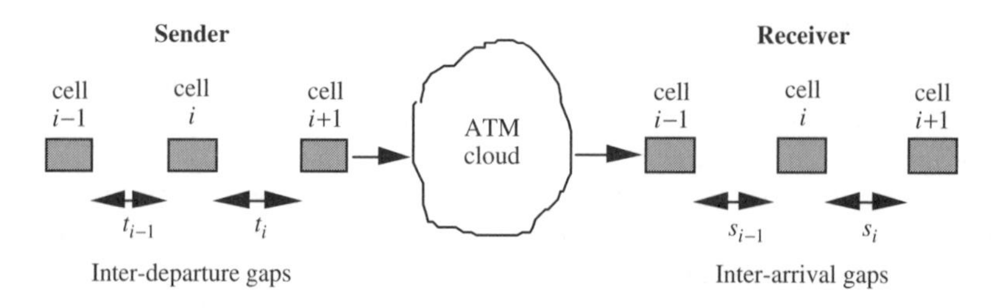

Figure 7.7 Inter-departure and inter-arrival gaps.

as shown in Figure 7.8. For example, let us assume that the max CTD is set to 20 ms and the fixed CTD is equal to 12 ms. Then, there is no guarantee that the peak-to-peak cell delay variation will always be less than 8 ms. The max CTD can be obtained as a percentile of the end-to-end cell transfer delay, so that the end-to-end cell transfer delay exceeds it only a small percent of the time. For instance, if it is set to the 99th percentile, then 99% of the time the end-to-end cell transfer delay will be less than max CTD, and 1% of the time it will be greater.

Of the quality-of-service parameters described above, the CLR, the peak-to-peak cell delay variation, and the max CTD have been standardized by the ATM Forum, and they can be signaled at call set-up time, i.e. at call set-up time, the calling party can specify values for these parameters. These values are upper bounds, and they represent the highest acceptable (and consequently the least desired) values. The values for the peak-to-peak cell delay variation and for the max CTD are expressed in msec. As an example, the calling party can request that the CLR is less or equal than 10^{-6}, the peak-to-peak cell delay variation is less or equal than 3 ms, and the max CTD is less or equal than 20 ms.

The network will accept the connection if it can guarantee the requested quality-of-service values. If it cannot guarantee these values, then it will reject the connection. Also, it is possible that the network and the calling party may negotiate new values for the quality-of-service parameters. As will be seen in the following section, the number of quality-of-service parameters signaled at call set-up time depends on the type of ATM service requested by the calling party.

Three additional quality-of-service parameters are used, namely the *Cell Error Rate* (CER), the *Severely Errored Cell Block Ratio* (SECBR) and the *Cell Misinsertion Rate* (CMR). These three parameters are not used by the calling party at call set-up. They are only monitored by the network.

The Cell Error Rate (CER) of a connection is the ratio of the number of *errored* cells, i.e. cells delivered to the destination with erroneous payload, to the total number of cells transmitted by the source.

The Severely Errored Cell Block Ratio (SECBR) is the ratio of the total number of severely errored *cell blocks* divided by the total number of transmitted cell blocks. A cell block is a sequence of cells transmitted consecutively on a given connection. A severely errored cell block occurs when more than a predefined number of errored cells, lost cells or misinserted cells are observed in a received cell block.

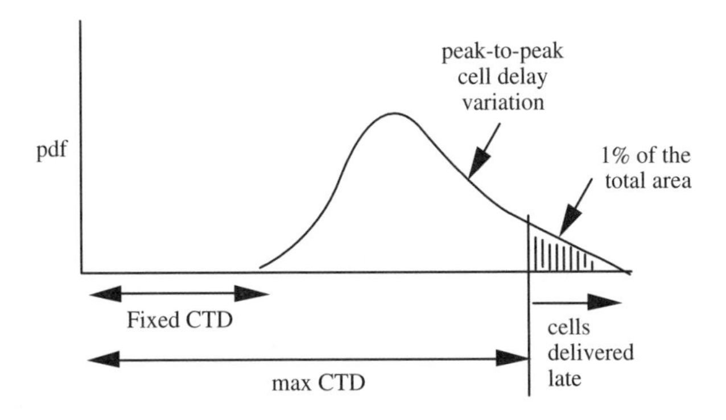

Figure 7.8 Cell transfer delay.

The Cell Misinsertion Rate (CMR) is the number of cells delivered to a wrong destination divided by a fixed time interval. A misinserted cell is a cell transmitted on a different connection due to an undetected error in its header.

7.3 ATM SERVICE CATEGORIES

An ATM service category is in simple terms a quality-of-service class. Each service category is associated with a set of traffic parameters and a set of quality-of-service parameters. Functions such as call admission control and bandwidth allocation (see Section 7.6) are applied differently for each service category. Also, as described in Section 6.7, the scheduling algorithm that determines in what order the cells in an output buffer of an ATM switch are transmitted out, provides different priorities to cells belonging to different service categories. In addition, a service category may be associated with a specific mechanism that is in place inside the network. The service category of a connection is signaled to the network at call set-up time, along with its traffic and quality-of-service parameters.

The ATM Forum has defined the following six service categories: *Constant Bit Rate* (CBR), *Real-Time Variable Bit Rate* (RT-VBR), *Non-Real-Time Variable Bit Rate* (NRT-VBR), *Available Bit Rate* (ABR), *Unspecified Bit Rate* (UBR) and *Guaranteed Frame Rate* (GFR). The first two service categories, namely CBR and RT-VBR, are for real-time applications, whereas the remaining service categories are for non-real-time applications.

The CBR Service

This service is intended for real-time applications which transmit at constant bit rate, such as circuit emulation services and constant-bit rate video.

Since the rate of transmission of a constant-bit rate application does not change over time, the peak cell rate is sufficient to describe the amount of traffic that the application transmits over the connection. The Cell Delay Variation Tolerance (CDVT) is also specified, and its use will be explained in Section 7.7.1. A CBR service is for real-time applications, and therefore the end-to-end delay is an important quality-of-service parameter. In view of this, in addition to the CLR, the two delay-related parameters, namely the peak-to-peak cell delay variation and the max CTD, are also specified.

In summary, the following traffic parameters are specified: PCR and CDVT. Also, the following quality-of-service parameters are specified: CLR, peak-to-peak cell delay variation, and max CTD.

The RT-VBR Service

This service is intended for real-time applications which transmit at a variable bit rate, such as encoded video and encoded voice.

Since the rate of transmission of a variable-bit rate application varies over time, the peak cell rate is not sufficient to describe the amount of traffic that the application will transmit over the connection. In addition to the PCR and the cell delay variation tolerance, the Sustained Cell Rate (SCR) and the Maximum Burst Size (MBS) are specified. As in the CBR service, the RT-VBR service is also intended for real-time applications. Therefore, in addition to the CLR, the two delay-related parameters, namely the peak-to-peak cell delay variation and the max CTD, are also specified.

In summary, the following traffic parameters are specified: PCR, CDVT, SCR, and MBS. Also, the following quality-of-service parameters are specified: CLR, peak-to-peak cell delay variation, and max CTD.

The NRT-VBR Service

This service is intended for non-real-time applications which transmit at a variable bit rate. As in the RT-VBR service, the traffic parameters PCR, the Cell Delay Variation Tolerance (CDVT), the Sustained Cell Rate (SCR) and the Maximum Burst Size (MBS) are specified. Since this service is not intended for real-time applications, only the CLR is specified.

In summary, the following traffic parameters are specified: PCR, CDVT, SCR, MBS. Also, the CLR quality-of-service parameter is specified.

The UBR Service

This is a best-effort type of service for non-real-time applications with variable bit rate. It is intended for applications that involve the transfer of data, such as file transfer, web browsing and email. No traffic or quality-of-service parameters are specified.

The PCR and the CDVT can be specified, but a network can ignore it. Also, a UBR user may indicate a *Desirable Minimum Cell Rate* (DMCR), but a network is not required to guarantee such as a minimum bandwidth.

The ABR Service

This service is intended for non-real-time applications which can vary their transmission rate according to the congestion level in the network.

A user requesting the ABR service specifies a *Minimum Cell Rate* (MCR), and a maximum cell rate, which is its PCR. The minimum cell rate could be zero. The user varies its transmission rate between its MCR and its PCR in response to feedback messages that it receives from the network. These feedback messages are conveyed to the user through a mechanism implemented in the network. During the time that the network has a slack capacity, the user is permitted to increase its transmission rate by an increment. When congestion begins to build up in the network, the user is requested to decrease its transmission rate by a decrement. A detailed description of the ABR service is given in Section 7.8.1.

The following traffic parameters are specified: PCR, CDVT and MCR. The CLR for ABR sources is expected to be low. Depending upon the network, a value for the CLR may be specified.

The GFR Service

This service is for non-real-time applications that require a Minimum Cell Rate (MCR) guarantee, but they can transmit in excess of their requested MCR. The application transmits data organized into frames, and the frames are carried in AAL 5 CPS-PDUs. The network does not guarantee delivery of the excess traffic. When congestion occurs, the network attempts to discard complete AAL 5 CPS-PDUs, rather than individual cells. The

GFR service does not provide explicit feedback to the user regarding the current level of congestion in the network. Rather, the user is supposed to determine network congestion through a mechanism such as TCP, and adapt its transmission rate.

The following traffic parameters are specified: PCR, Minimum Cell Rate (MCR), Maximum Burst Size (MBS), and *Maximum Frame Size* (MFS). The CLR for the frames that are eligible for the service guarantee is expected to be low. Depending upon the network, a value for the CLR may be specified.

ATM Transfer Capabilities

In the ITU-T standard, the ATM service categories are referred to as *ATM transfer capabilities.* Some of the ATM transfer capabilities are equivalent to ATM Forum's service categories, but they have a different name. The CBR service is called the *Deterministic Bit Rate* (DBR) service, the RT-VBR service is called the *Real-Time Statistical Bit Rate* (RT-SBR) service and the NRT-VBR service is called the *Non-Real-Time Statistical Bit Rate* (NRT-SBR) service. The UBR service category has no equivalent ATM transfer capability. Both the ABR and GFR services have been standardized by ITU-T. Finally, the ITU-T ATM transfer capability *ATM Block Transfer* (ABT), described in Section 7.6.2, has no equivalent service category in the ATM Forum standards.

7.4 CONGESTION CONTROL

Congestion control procedures can be grouped into the following two categories: *preventive* and *reactive* control.

In preventive congestion control, as its name implies, we attempt to prevent congestion from occurring. This is achieved using the following two procedures: *Call (or connection) Admission Control* (CAC) and *bandwidth enforcement.* Call admission control is exercised at the connection level, and it is used to decide whether to accept or reject a new connection. Once a new connection has been accepted, bandwidth enforcement is exercised at the cell level to assure that the source transmitting on this connection is within its negotiated traffic parameters.

Reactive congestion control is based on a different philosophy than preventive congestion control. In reactive congestion control, the network uses feedback messages to control the amount of traffic that an end device transmits so that congestion does not arise.

In the remaining sections of the chapter, we examine in detail various preventive and reactive congestion control schemes.

7.5 PREVENTIVE CONGESTION CONTROL

As mentioned above, preventive congestion control involves the following two procedures: Call Admission Control (CAC) and bandwidth enforcement. Call admission control is used by the network to decide whether to accept a new connection or not.

As we have seen so far, ATM connections maybe either Permanent Virtual Connections (PVC) or Switched Virtual Connections (SVC). A PVC is established manually by a network administrator using network management procedures, whereas an SVC is established in real-time by the network using the signaling procedures described in Chapters 10 and 11.

In our discussion below, we will consider a point-to-point SVC. We recall that point-to-point connections are bidirectional. The traffic and quality of service parameters can be different for each direction of the connection.

Let us assume that an end device, referred to as end device 1, wants to set-up a connection to a destination end device, referred to as end device 2. A point-to-point SVC is established between the two end devices, when end device 1 sends a SETUP message to its ingress switch, referred to as switch A, requesting that a connection is established to end device 2. The ingress switch calculates a path through the network to the switch to which the destination end device is attached using a routing algorithm. We shall refer to this switch as B. Then, it forwards the set-up request to its next-hop switch, which in turn forwards it to its next-hop switch, and so on until it reaches switch B. Switch B sends the set-up request to end device 2, and if it is accepted, a confirmation message is sent back to end device 1.

The set-up message, as will be seen in Chapter 10, contains a variety of different types of information, including values for the traffic and quality-of-service parameters. This information is used by each switch in the path to decide whether it should accept or reject the new connection. This decision is based on the following two questions:

1. Will the new connection affect the quality-of-service of the existing connections already carried by the switch?
2. Can the switch provide the quality-of-service requested by the new connection?

As an example, let us consider a non-blocking ATM switch with output buffering, shown in Figure 6.23, and let us assume that the quality of service is measured by the cell loss rate. Typically, the traffic that a specific output port sees is a mixture of different connections that enter the switch from different input ports. We assume that so far the switch provides a cell loss probability of 10^{-6} for each existing connection routed through this output port. Now, let us assume that the new connection requests a cell loss rate of 10^{-6}. What the switch has to decide is whether the new cell loss rate for both the existing connections and the new connection will be 10^{-6}. If the answer is yes, then the switch can accept the new connection. Otherwise it will reject it.

Each switch on the path of a new connection has to decide, independently of the other switches, whether it has enough bandwidth to provide the quality-of-service requested for this connection. This is done using a CAC algorithm. Various CAC algorithms are discussed in the following section.

If a switch along the path is unable to accept the new connection, then the switch refuses the set-up request, and it sends it back to a switch in the path that can calculate an alternative path.

Once a new connection has been accepted, bandwidth enforcement is exercised at the cell level to assure that the transmitting source is within its negotiated traffic parameters. Bandwidth enforcement procedures are discussed in Section 7.7.

7.6 CALL ADMISSION CONTROL (CAC)

Call Admission Control (CAC) algorithms may be classified into *non-statistical bandwidth allocation* (or *peak bit rate allocation*) and *statistical bandwidth allocation*. Below, we examine these two classes.

Non-Statistical Bandwidth Allocation

Non-statistical bandwidth allocation, otherwise known as peak bit rate allocation, is used for connections requesting a CBR service. In this case, the CAC algorithm is very simple, as the decision to accept or reject a new connection is based purely on whether its peak bit rate is less than the available bandwidth on the link. Let us consider, for example, a nonblocking switch with output buffering, such as the one shown in Figure 6.23, and suppose that a new connection with a peak bit rate of 1 Mbps has to be established through output link 1. Then, the new connection is accepted if the link's available capacity is more or equal to 1 Mbps.

In the case where nonstatistical allocation is used for all the connections routed through a link, the sum of the peak bit rates of all the existing connections is less than the link's capacity. Peak bit rate allocation may lead to a grossly under-utilized link, unless the connections transmit continuously at peak bit rate.

Statistical Bandwidth Allocation

In statistical bandwidth allocation, the allocated bandwidth on the output link is less than the peak bit rate of the source. In the case where statistical allocation is used for all the connections on the link, the sum of the peak bit rates of all the connections may exceed the link's capacity. Statistical allocation makes economic sense when dealing with bursty sources, but it is difficult to implement effectively. This is due to the fact that it is not always possible to characterize accurately the traffic generated by a source and how it is modified deep in an ATM network. For instance, let us assume that a source has a maximum burst size of 100 cells. As the cells belonging to the same burst travel through the network, they get buffered in each switch and due to multiplexing with cells from other connections and scheduling priorities, the maximum burst of 100 cells may become much larger deep in the network. Other traffic descriptors, such as the PCR and the SCR, can be similarly modified deep in the network. For instance, let us consider a source with a peak bit rate of 128 Kbps. Due to multiplexing and scheduling priorities, it is possible that several cells from this source can get batched together in the buffer of an output port of a switch. Now, let us assume that this output port has a speed of, say 1.544 Mbps. Then, these cells will be transmitted out back-to-back at 1.544 Mbps, which will cause the peak bit rate of the source to increase temporarily!

Another difficulty in designing a CAC algorithm for statistical allocation is due to the fact that an SVC has to be set-up in real-time. Therefore, the CAC algorithm cannot be CPU intensive. This problem may not be as important when setting up PVCs. The problem of whether to accept or reject a new connection may be formulated as a queueing problem. For instance, let us consider again our nonblocking switch with output buffering. The CAC algorithm has to be applied to each output port. If we isolate an output port and its buffer from the switch, we will obtain the queueing model shown in Figure 7.9. This type of queueing structure is known as the *ATM multiplexer*. It represents a number of ATM sources feeding a finite-capacity queue which is served by a server, i.e. the output port. The service time is constant and it is equal to the time it takes to transmit an ATM cell. (The simulation project described at the end of this chapter deals with an ATM multiplexer.)

Now let us assume that the quality of service, expressed in cell loss rate, of the existing connections is satisfied. The question that arises is whether the cell loss rate will still

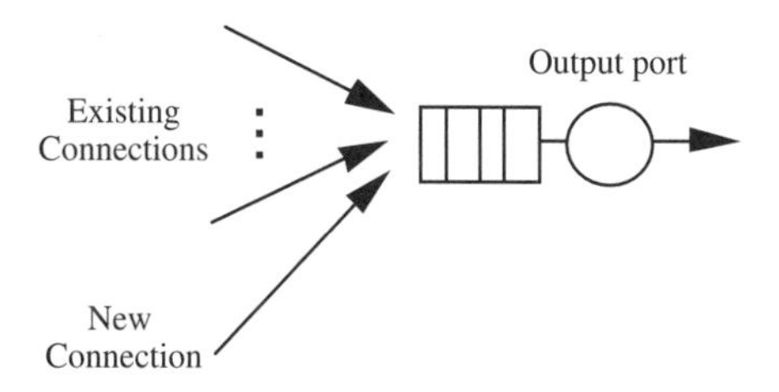

Figure 7.9 An ATM multiplexer.

be maintained if the new connection is accepted. This can be answered by solving the ATM multiplexer queueing model with the existing connections and the new connection. However, the solution to this problem is CPU intensive, and it cannot be done in real-time. In view of this, a variety of different call admission control algorithms have been proposed which do not require the solution of such a queueing model.

Most of the call admission control algorithms that have been proposed are based solely on the cell loss rate QoS parameter. That is, the decision to accept or reject a new connection is based on whether the switch can provide the new connection with the requested cell loss rate without affecting the cell loss rate of the existing connections. No other QoS parameters, such as peak-to-peak cell delay variation and the max CTD, are considered by these algorithms. A very popular example of this type of algorithm is the *equivalent bandwidth*, described below.

CAC algorithms based on the cell transfer delay have also been proposed. In these algorithms, the decision to accept or reject a new connection is based on a calculated absolute upper bound of the end-to-end delay of a cell. These algorithms are closely associated with specific scheduling mechanisms, such as static priorities, early-deadline-first, and weighted fair queueing. Given that the same scheduling algorithm runs on all the switches in the path of a connection, it is possible to construct an upper bound of the end-to-end delay. If this is less than the requested end-to-end delay, then the new connection is accepted.

Below, we examine the equivalent bandwidth scheme and then we present the ATM Block Transfer (ABT) scheme used for bursty sources. In this scheme, bandwidth is allocated on demand and only for the duration of a burst. Finally, we present a scheme for controlling the amount of traffic in an ATM network based on Virtual Path Connections (VPC).

7.6.1 Equivalent Bandwidth

Let us consider a finite capacity queue served by a server at the rate of μ. This queue can be seen as representing an output port and its buffer in a nonblocking switch with output buffering. We assume that this queue is fed by a single source, whose equivalent bandwidth we wish to calculate. Now, if we set μ equal to the source's peak bit rate, then we will observe no accumulation of cells in the buffer. This is because the cells arrive as fast as they are transmitted out. Now, if we slightly reduce the service rate μ, then we will see that cells are beginning to accumulate in the buffer. If we reduce the service rate a little bit more, then the buffer occupancy will increase. If we keep repeating this experiment and each time we lower slightly the service rate, then we will see that the cell loss rate begins to increase. The equivalent bandwidth of the source is defined as the service rate e at which the queue is served that corresponds to a cell loss rate of ε. The equivalent bandwidth of a source falls somewhere between its average bit rate and its peak bit rate. If the source is very bursty, it is closer to its peak bit rate, otherwise it is closer to its average bit rate. We note that the equivalent bandwidth of a source is not related the source's SCR.

There are various approximations that can be used to compute quickly the equivalent bandwidth of a source. A commonly used approximation is based on the assumption that the source is an Interrupted Fluid Process (IFP) characterized by the triplet (R,r,b), where R is its peak bit rate, r the fraction of time the source is active, defined as the ratio of the mean length of the on period divided by the sum of the mean on and off periods, and b is the mean duration of the on period. Let us now assume that the source feeds a finite-capacity queue with a constant service time, and let K be the size of the queue expressed in bits. The service time is equal to the time it takes to transmit out a cell. Then, the equivalent bandwidth e is given by the expression:

$$e = \frac{a - K + \sqrt{(a-K)^2 + 4Kar}}{2a} R \tag{7.1}$$

where $a = b(1-r)R \ln(1/\varepsilon)$.

The equivalent bandwidth of a source is used in statistical bandwidth allocation in the same way that the peak bit rate is used in nonstatistical bandwidth allocation. For instance, let us consider an output link of a nonblocking switch with output buffering, and let us assume that it has a transmission speed of 25 Mbps and its associated buffer has a capacity of 200 cells. We assume that no connections are currently routed through the link. The first set-up request that arrives is for a connection that requires an equivalent bandwidth of 5 Mbps. The connection is accepted and the link has now 20 Mbps available. The second set-up request arrives during the time that the first connection is still up, and it is for a connection that requires 10 Mbps. The connection is accepted and 10 Mbps are reserved, leaving 10 Mbps free. If the next set-up request arrives during the time that the first two connection are still up and it is for a connection that requires more than 10 Mbps, then the new connection is rejected.

This method of simply adding up the equivalent bandwidth requested by each connection may lead to under-utilization of the link, i.e. more bandwidth may be allocated for all the connections than it is necessary. The following approximation for the equivalent bandwidth of N sources corrects the over-allocation problem:

$$c = \min\left\{\rho + \sigma\sqrt{-2\ln(\varepsilon) - \ln(2\pi)}, \sum_{i=1}^{N} e_i\right\} \tag{7.2}$$

where ρ is the average bit rate of all the sources, e_i is the equivalent bandwidth of the ith source, calculated using expression (7.1), and σ is the sum of the standard deviation of the bit rate of all the sources, and it is equal to

$$\sigma = \sum_{i=1}^{N} \sqrt{r_i(R_i - r_i)}$$

When a new set-up request arrives, the equivalent bandwidth for all the existing connections and the new one is calculated using expression (7.2). The new connection is accepted if the resulting bandwidth c is less than the link's capacity.

Below we demonstrate through a numerical example how the maximum number of connections admitted using the above expressions for the equivalent bandwidth varies with the buffer size K, the cell loss rate ε, and the fraction of time the source is active r. We consider a link with a transmission speed C equal to 150 Mbps and a buffer capacity of K cells. Each connection is characterized by the parameters R, its peak bit rate, ρ the average bit rate, and b the mean duration of the on period. We note that the quantity r defined above is related to ρ through the expression $rR = \rho$. In the numerical examples presented below, we assume that all connections are identical with traffic parameters $(R,\rho,b) = (10\,\text{Mbps}, 1\,\text{Mbps}, 310\,\text{cells})$.

We note that if we admit connections using their peak bit rate, then a maximum of $150\,\text{Mbps}/10\,\text{Mbps} = 15$ connections will be admitted. On the other hand, if we admit connections using the average bit rate, a maximum of $150\,\text{Mbps}/1\,\text{Mbps} = 150$ connections will be admitted. These two values can be seen as an upper and lower bound on the number of connections that can be admitted using the equivalent bandwidth method.

In Figure 7.10, the maximum number of connections that can be admitted is plotted as a function of the buffer size K. The buffer size was increased from 31 cells to 31 000 cells. For each value of K, the maximum number of admitted connections was obtained using expressions (7.1) and (7.2) with the cell loss rate fixed to 10^{-6}. We observe that for small values of K, the maximum number of connections admitted by the equivalent bandwidth algorithm is constant. As K increases, the maximum number of admitted connections increases as well, and eventually flattens out.

In Figure 7.11, the maximum number of admitted connections is plotted against the cell loss rate ε. The buffer size was fixed to 1236 cells, i.e. 64 Kbytes. We observe that the maximum number of admitted connections is not very sensitive to the cell loss rate ε. In this particular example, the buffer size is large enough so that the equivalent algorithm admits a large number of connections. In general, the equivalent bandwidth algorithm becomes more sensitive to ε when the buffer size is smaller.

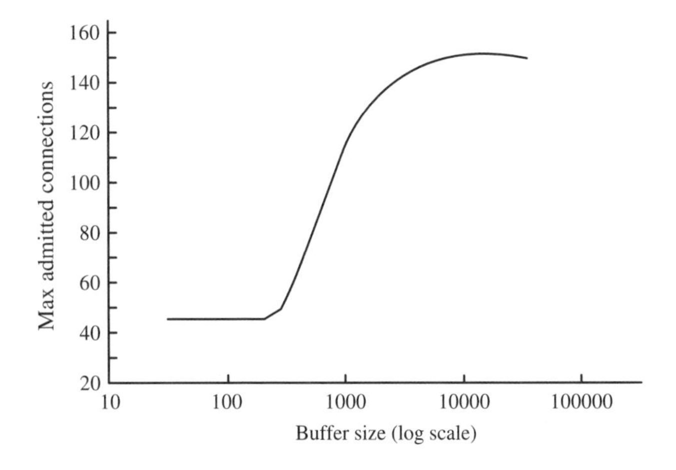

Figure 7.10 Varying the buffer size K.

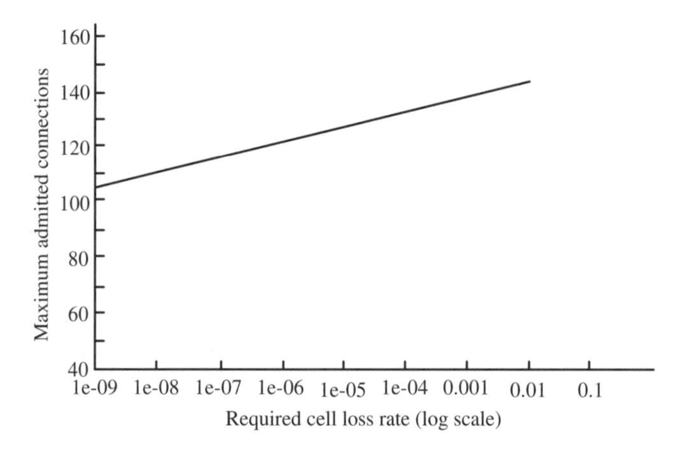

Figure 7.11 Varying the required cell loss rate ε.

Finally, in Figure 7.12, the maximum number of admitted connections is plotted against r, the fraction of time that a source is active, where $r = \rho/R$. We recall from Section 7.1.2, that r can be used to express the burstiness of a source. The buffer size was fixed to 1236 cells and the cell loss rate ε to 10^{-6}. We observe that the maximum number of admitted connections dependents on r. As r increases, the source becomes more bursty and requires more buffer space in order to maintain the same cell loss rate. As a result the maximum number of admitted connections falls sharply as r tends to 0.5.

7.6.2 The ATM Block Transfer (ABT) Scheme

A number of congestion control schemes were devised for bursty sources whereby each switch allocates bandwidth on demand, and only for the duration of a burst. The main

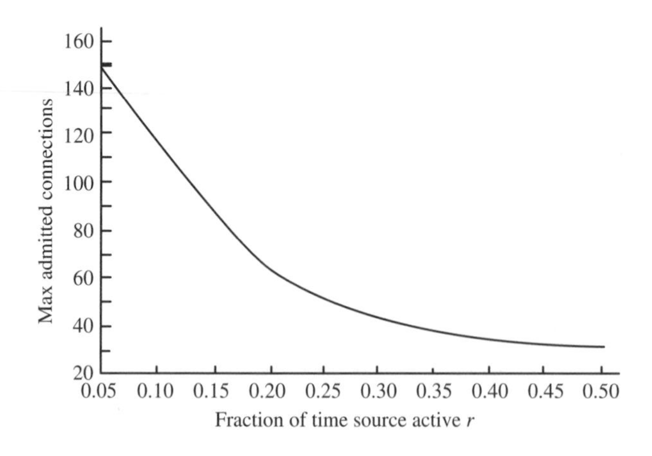

Figure 7.12 Varying r.

idea behind these schemes is the following. At connection set-up time, the path through the ATM network is selected, and each switch in the path allocates the necessary VPI/VCI labels and updates the switching table used for label swapping in the usual way. However, it does not allocate any bandwidth to this connection. When the source is ready to transmit a burst, it notifies the switches along the path, and it is at that moment that each switch will allocate the necessary bandwidth for the duration of the burst.

These congestion control schemes are known as *fast bandwidth allocation* schemes. The ATM Block Transfer (ABT) scheme is a fast bandwidth allocation scheme, and it is a standardized ATM transfer capability. ABT makes use only of the peak bit rate, and it is intended for VBR sources whose peak bit rate is less than 2% of the link's capacity.

In ABT, a source requests bandwidth in incremental and decremental steps. The total requested bandwidth for each connection may vary between zero and its peak bit rate. For a step increase, a source uses a special reservation request cell. If the requested increase is accepted by all the switches in the path, then the source can transmit at the higher bit rate. If the step increase is denied by a switch in the path, then the step increase request is rejected. Step decreases are announced through a management cell. A step decrease is always accepted. At the cell level, the incoming cell stream of a source is shaped, so that the enforced peak bit rate corresponds to the currently accepted peak bit rate.

A Fast Reservation Protocol (FRP) unit was implemented to handle the relevant management cells. This unit is located at the UNI. The protocol utilizes different timers to ensure its reliable operation. The end device utilizes a timer to ensure that its management cells, such as step increase requests, sent to its local FRP unit are not lost. When the FRP unit receives a step increase request, it forwards the request to the first switch in the path, which in turn forwards it to the next-hop switch, and so on. If the request can be satisfied by all the switches on the path, then the last switch will send an ACK to the FRP unit. The FRP unit then informs the end device that the request has been accepted, updates the policing function, and sends a validation cell to the switches in the path to confirm the reservation. If the request cannot be satisfied by a switch, the switch simply discards the request. The upstream switches, which have already reserved bandwidth, will discard the reservation if they do not receive the validation cell by the time a timer expires. This timer is set equal to the maximum round trip delay between the FRP unit and the furthermost switch. If the request is blocked, the FRP unit will retry to request the step increase after a period set by another timer. The number of attempts is limited.

This mechanism can be used by an end device to transmit bursts. When the end device is ready to transmit a burst, it issues a step increase request with a requested bandwidth equal to its peak bit rate. If the request is granted, the end device transmits its burst, and at the end it announces a step decrease with bandwidth equal to its peak bit rate.

In a slightly different version of the ABT protocol, the end device starts transmitting its burst immediately after it issues a reservation request. The advantage of this scheme is that the end device does not have to wait until the request is granted. The burst will get lost if a switch in the path is unable to accommodate the request.

7.6.3 Virtual Path Connections

A virtual path connection can be used in an ATM network to create a dedicated connection between two switches. Within this connection, individual virtual circuit connections can be set up without the knowledge of the network.

Let us assume, for instance, that a permanent virtual path connection is established between two switches, namely 1 and 2. These two switches may not be adjacent, and they may communicate through several other switches. A fixed amount of bandwidth is allocated to the virtual path connection. This bandwidth is reserved for this particular connection, and it cannot be shared with other connections, even when it is not used entirely. An end device attached to switch 1 and wishing to communicate to an end device attached to switch 2 is allocated part of the bandwidth of the virtual path connection using nonstatistical or statistical bandwidth allocation. The connection is rejected if there is not enough bandwidth available within the virtual path connection, since the total amount of traffic carried by this virtual path connection cannot exceed its allocated bandwidth.

A virtual channel connection maintains the same VCI value through out an entire virtual path connection, i.e. its VCI value has global significance. The virtual path, however, is identified by a series of VPI values, each having local significance.

An example of label swapping in a virtual path connection is given in Figure 7.13. A virtual path connection has been established between switches 1 and 3. Users A, B and C are attached to switch 1, and via the virtual path connection, they are connected to their respective destinations A′, B′ and C′, which are attached to switch 3. Each switch is represented by a square, and its switching table is given immediately below the square. We assume that the switching table is centralized, and it contains information for all input ports. The first three columns in the switching table give the VPI/VCI of each incoming connection and its input port. The second three columns give the new label and the destination output port of the connection. Also, the first row of each switching table is associated with the connection from A to A′, the second row is associated with the connection from B to B′, and the third row is associated with the connection from C to C′. We observe, that the virtual path connection has a VPI = 1 on the UNI between users A, B, C and switch 1, a VPI = 5 on the hop from switch 1 to switch 2, a VPI = 6 on the hop from switch 2 to switch 3, and a VPI = 7 on the UNI between switch 3 and users A′, B′ and C′. The virtual channel connections from A to A′, B to B′ and C to C′ are identified by the VCIs 47, 39 and 41, respectively.

Virtual path connections provide a network operator with a useful mechanism. For instance, it can be used to provide a customer with a dedicated connection between two locations. Within this connection, the customer can set up any number of virtual circuit connections, as long as the total bandwidth allocated to the virtual path connection is not exceeded.

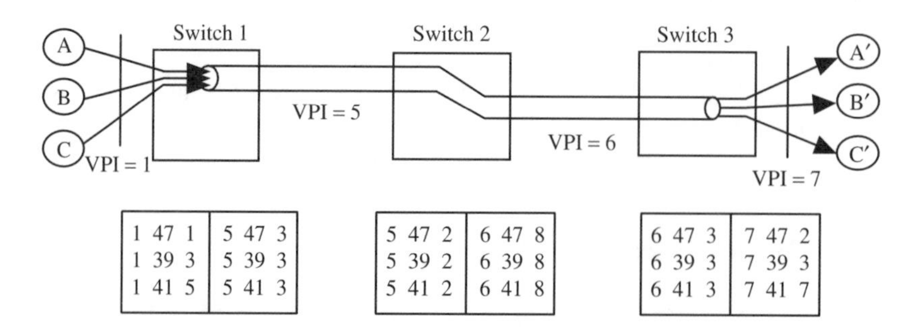

Switch 1:

1	47	1	5	47	3
1	39	3	5	39	3
1	41	5	5	41	3

Switch 2:

5	47	2	6	47	8
5	39	2	6	39	8
5	41	2	6	41	8

Switch 3:

6	47	3	7	47	2
6	39	3	7	39	3
6	41	3	7	41	7

Figure 7.13 Label swapping in a virtual path connection.

Virtual path connections can be combined to form a virtual network overlaid on an ATM network. Such a virtual network can be set up by a network operator to control the amount of traffic in the network. In addition, the network operator can set up different virtual networks for different ATM service categories.

7.7 BANDWIDTH ENFORCEMENT

The function of bandwidth enforcement is to ensure that the traffic generated by a source conforms with the *traffic contract* that was agreed upon between the user and the network at call set-up time. According to the ITU-T and the ATM Forum, the traffic contract consists of (1) the traffic parameters, (2) the requested quality of service parameters, and (3) a definition of conformance. The traffic and the quality-of-service parameters, as we have seen, depend upon the service category requested.

Testing the conformance of a source, otherwise known as policing the source, is carried out at the User-Network Interface (UNI). It involves policing the peak cell rate and the sustained cell rate using the Generic Cell Rate Algorithm (GCRA). ITU-T first standardized this algorithm for the peak cell rate. The ATM Forum adapted the same algorithm, and it also extended it for testing the conformance of the sustained cell rate. It is possible that multiple GCRAs can be used in series, such as one for the peak cell rate and another one for the sustained cell rate.

Policing each source is an important function from the point of view of a network operator, since a source exceeding its contract may affect the quality-of-service of other existing connections. Also, depending upon the pricing scheme used by the network operator, there may be loss of revenue. A source may exceed its contract due to various reasons, such as intentional or unintentional under-estimation by the user of the required bandwidth, and malfunctioning of a user's equipment.

The generic cell rate algorithm is based on a popular policing mechanism known as the *leaky bucket*. The leaky bucket may be *unbuffered* or *buffered*. The unbuffered leaky bucket consists of a token pool of size K, as shown in Figure 7.14(a). Tokens are generated at a fixed rate. A token is lost if it is generated at a time when the token pool is full. An arriving cell takes a token from the token pool, and then enters the network. The number of tokens in the token pool is then reduced by one. A cell is considered to be a *violating* cell (or, a *noncompliant* cell), if it arrives at a time when the token pool is empty. The buffered leaky bucket is shown in Figure 7.14(b). It is the same as the unbuffered leaky bucket with the addition of an input buffer of size M, where a cell can wait if it arrives at a time when the token pool is empty. A cell is considered to be a violating cell if it arrives at a time when the input buffer is full. Violating cells are either dropped or tagged (see Section 7.7.2).

A leaky bucket is completely defined by its parameters: K, token generation rate, and M, if it is a buffered leaky bucket. The difficulty with the leaky bucket is in fixing its parameters, so that it is transparent when the source adheres to its contract, and it catches all the violating cells when the source exceeds its contract. Given a probabilistic model of an arrival process of cells to the UNI, it is possible to fix the parameters of the leaky bucket using queueing-based models. However, it has been shown that the leaky bucket can be very ineffective in catching violating cells. Dual leaky buckets have been suggested for more efficient policing, where the first leaky bucket polices violations of the peak cell rate, and the second one polices violations of the source's burstiness. As will be seen

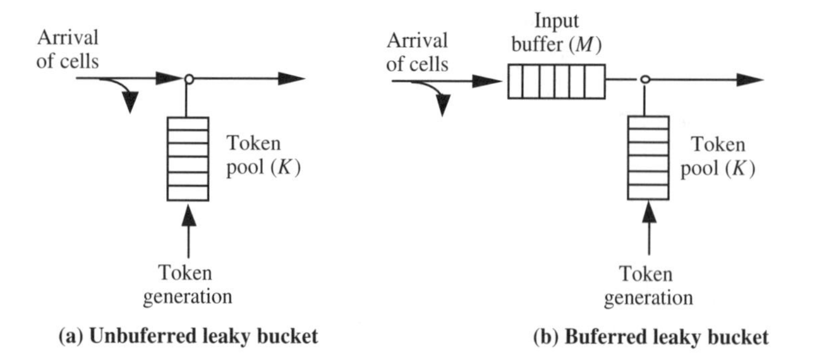

Figure 7.14 The leaky bucket.

below, GCRA does catch all violating cells, but to do that it needs an additional traffic parameter.

In addition to GCRA, a source can shape its traffic using a *traffic shaper* to attain desired characteristics for the stream of cells it transmits to the network. Traffic shaping involves peak cell rate reduction, burst size reduction, and reduction of cell clumping by suitably spacing out the cells in time.

7.7.1 The Generic Cell Rate Algorithm (GCRA)

Unlike the leaky bucket mechanism, GCRA is a deterministic algorithm and it does catch all the violating cells. However, for this it requires an additional new traffic parameter known as the Cell Delay Variation Tolerance (CDVT). This parameter is not to be confused with the peak-to-peak cell delay variation parameter described in Section 7.2.

Let us assume that a source is transmitting at peak cell rate and it produces a cell every T units of time, where $T = 1/\text{PCR}$. As shown in Figure 7.15, due to multiplexing with cells from other sources and with signaling and network management cells, it is possible that the inter-arrival time of successive cells belonging to the same source at the UNI may vary around T. That is, for some cells it may be greater than T, and for others it may be less than T. In the former case, there is no penalty in arriving late! However, in the latter case, the cells will appear to the UNI that they were transmitted at a higher rate, even though they were transmitted conformally to the peak cell rate. In this case, these cells should not be penalized by the network. The cell delay variation tolerance is a parameter that permits the network to tolerate a number of cells arriving at a rate which is faster

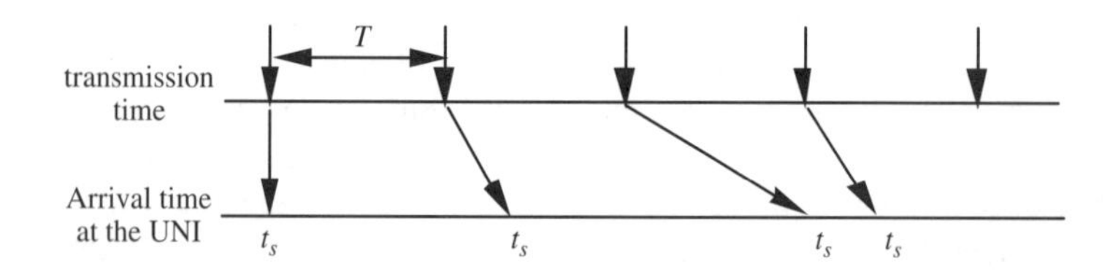

Figure 7.15 Arrival times at the UNI.

than the agreed upon peak cell rate. This parameter does not depend upon a particular source, rather it depends on the number of sources that use the same UNI and the access to the UNI, and it is specified by a network administrator.

GCRA can be used to monitor the peak cell rate and the sustained cell rate. There are two implementations of GCRA, namely, the *virtual scheduling algorithm* and the *continuous-state leaky bucket* algorithm. These two algorithms are equivalent to each other.

Policing The Peak Cell Rate

To monitor the peak cell rate, the following two parameters are required: peak emission interval T and cell delay variation tolerance τ. $T = 1/\text{PCR}$, and it is obtained from the user's declared peak cell rate, and as mentioned above, τ is provided by a network administrator.

A flowchart of the virtual scheduling algorithm is shown in Figure 7.16. Variable TAT is the theoretical arrival time of a cell, and t_s is the actual arrival time of a cell. At the time of arrival of the first cell, TAT $= t_s$. Each time a cell arrives, the algorithm calculates the theoretical time TAT of the next arrival. If the next cell arrives late, that is if TAT $< t_s$, then the next theoretical arrival time is set to TAT $= t_s + T$. If the next arrival is early, that is TAT $> t_s$, then the cell may be accepted or it may be classified as noncompliant. The decision is based on the cell delay variation tolerance τ and also on previously arrived cells that were late but they were accepted as compliant. Specifically, if TAT $< t_s + \tau$, then the cell is considered as compliant. Notice, however, that the next theoretical arrival time TAT is set equal to the theoretical arrival time of the current cell plus T, that is TAT $=$ TAT $+ T$. If the next arrival occurs before the theoretical arrival

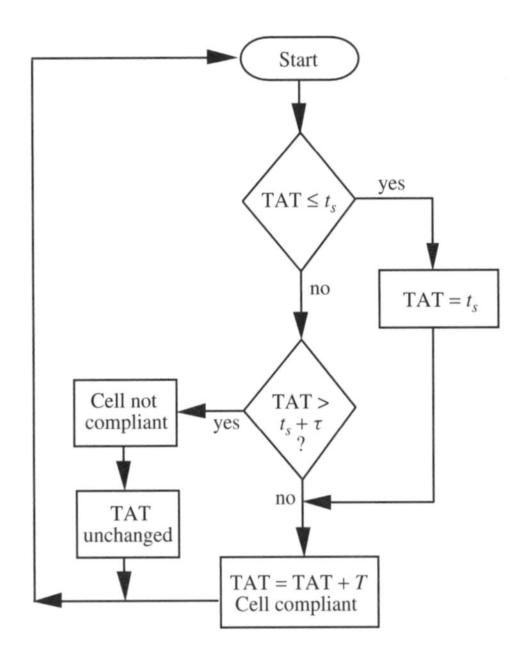

Figure 7.16 The virtual scheduling algorithm.

time TAT, it may still be accepted if TAT $< t_s + \tau$. However, if cells continue to arrive early, the cell delay variation will be used up, and eventually a cell will be classified as nonconformant.

As an example, let us consider the case where $T = 10$, $\tau = 15$, and the actual arrival times of the first five cells are: 0, 12, 18, 20 and 25. For cell 1 we have that $t_s =$ TAT $= 0$. The cell is accepted and TAT is set to TAT $+ 10 = 10$. For cell 2, $t_s = 12$, and since TAT $\leq t_s$, the cell is accepted and TAT is set equal to $t_s + T = 22$. Cell 3 arrives at time $t_s = 18$, and in view of this, TAT $> t_s$. Since TAT $\leq t_s + \tau$ the cell is accepted, and TAT is set equal to TAT $+ T = 32$. Cell 4 arrives at time $t_s = 20$, and TAT $> t_s$. Since TAT $\leq t_s + \tau$, the cell is accepted, and TAT is set equal to TAT $+ T = 42$. Cell 5 is not as lucky as cells 3 and 4. Its arrival time is $t_s = 25$ which makes TAT $> t_s$. Since TAT $> t_s + \tau$ the cell is considered as noncompliant.

A flowchart of the continuous state leaky bucket algorithm is shown in Figure 7.17. In this algorithm, a finite-capacity leaky bucket is implemented whose real-value content is drained out at a continuous rate of 1 unit of content per unit-time. Its content is increased by a fixed increment T each time a conforming cell arrives. The algorithm makes use of the variables X, X' and LCT. X indicates the current value of the leaky bucket, X' is an auxiliary variable, and LCT is the Last Compliance Time, that is, the last time a compliant cell arrived. At the arrival time of the first cell, $X = 0$ and $LCT = t_s$.

When a cell arrives, the quantity $X - (t_s - LCT)$ is calculated and saved in the auxiliary variable X'. If $X' \leq 0$, then the cell has arrived late and the cell is accepted, X is increased by T, and $LCT = t_s$. If $X' > 0$, then depending upon whether X' is less or greater than τ, the cell is accepted or it is considered as noncompliant. If $X > \tau$, the cell

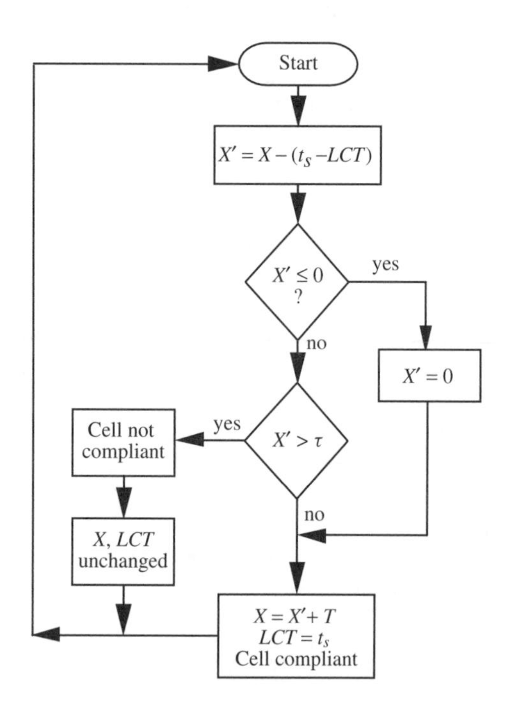

Figure 7.17 Continuous state leaky bucket algorithm.

is classified as noncompliant and the values of X and LCT remain unchanged. If $X \leq \tau$, then the cell is accepted, X is set to $X' + T$, and $LCT = t_s$.

Let us now consider the same example as in the virtual scheduling algorithm, that is $T = 10$, $\tau = 15$, and the actual arrival times of the first five cells are: 0, 12, 18, 20 and 25. For cell 1 we have that $X = 0$ and $LCT = 0$. The cell is accepted and X is set to 10 and LCT to 0. For cell 2, we have $X' = -2$. The cell is accepted and X is set to 10 and LCT to 12. Cell 3 arrives at time 18, which gives a value for X' equal to 4. Since $X' < \tau$, the cell is accepted and X is set to $X' + T = 14$ and LCT to 18. Cell 4 arrives at time 20, and we have $X' = 12$. Since $X' < \tau$, the cell is accepted and X is set to $X' + T = 22$ and LCT to 20. Cell 5 arrives at time 25, and we have that $X' = 17$. Since $X' > \tau$, the cell is classified as noncompliant and X and LCT remain unchanged.

Policing The Sustained Cell Rate

The sustained cell rate of a source is policed by either GCRA algorithm. As we saw above, GCRA uses the parameters T and τ to police the peak cell rate of a source. For policing the sustained cell rate of a source, it uses the parameters T_s and τ_s. T_s is the emission interval when the source transmits at its sustained cell rate, and it is equal to 1/SCR. τ_s is known as the *Burst Tolerance* (BT), and it is calculated from the Maximum Burst Size (MBS) provided by the source using the expression:

$$\tau_s = (MBS - 1)(T_s - T)$$

If the inter-arrival time of cells is equal to or greater than T_s, then the cells are compliant. However, some cells may arrive every T units of time, where $T < T_s$, if they are transmitted at peak cell rate. Since these cells arrive every T units of time, they are in essence noncompliant as far as GCRA is concerned. How many such cells should GCRA tolerate, before it starts classifying them as noncompliant? Obviously, the maximum number of cells that can arrive every T units of time is equal to the source's MBS minus the first cell that initiates the burst. That is, we expect a maximum of (MBS – 1) cells to arrive $(T_s - T)$ units of time faster. This gives a total time of (MBS – 1)$(T_s - T)$, which is the burst tolerance τ_s. In conformance testing, τ_s is set equal to:

$$\tau_s = (MBS - 1)(T_s - T) + CDVT$$

7.7.2 Packet Discard Schemes

As we saw in the previous section, GCRA will either accept a cell or classify it as noncompliant. Therefore, the question that arises is what to do with noncompliant cells. The simplest scheme is to just drop them. A more popular mechanism, known as *violation tagging*, attempts to carry the noncompliant cells if there is slack capacity in the network. The violating cells are tagged at the UNI, and then they are allowed to enter the network. If congestion arises inside the network, the tagged cells are dropped. Tagging of a cell is done using the Cell Loss Priority (CLP) bit in the cell's header. If the cell is untagged, then its CLP = 0. When a cell is tagged, its CLP = 1.

Violation tagging introduces two types of cells: the untagged and tagged cells. A simple way to handle tagged cells is through a priority mechanism, such as the *push-out* or *threshold* schemes. In the push-out scheme, both untagged and tagged cells are freely admitted into a buffer as long as the buffer is not full. If a tagged cell arrives during the time that the buffer is full, the cell is lost. If an untagged cell arrives during the time that the buffer is full, the cell will take the space of the last arrived tagged cell. The untagged cell will get lost if all the cells in the buffer are untagged. In the threshold scheme, both untagged and tagged cells are admitted as long as the total number of cells is below a threshold. Over the threshold, only untagged cells are admitted, and the tagged cells are rejected. The push-out priority scheme is more efficient than the threshold priority scheme, but the latter is preferable because it is simpler to implement. Other priority mechanisms have also been proposed, such as dropping from the front. This mechanism is similar to the threshold mechanism, only cells are dropped from the front. That is, when a tagged cell is ready to begin its service, the total number of cells in the buffer is compared against the threshold. If it is below, service begins, else the cell is dropped.

A discarded cell may be part of a user packet, such as a TCP packet. In this case, the receiving TCP will detect that the packet is corrupted, and it will request the sending TCP to retransmit it. In view of this, when discarding a cell we can save bandwidth by discarding the subsequent cells that belong to the same user packet since the entire packet will have to be retransmitted anyway. For applications using AAL 5, it is possible to identify the beginning and the end of each user packet, and consequently drop the subsequent cells that belong to the same packet. There are two such discard mechanisms, namely *Partial Packet Discard* (PPD) and *Early Packet Discard* (EPD). Partial packet discard can be applied when the discarded cell is not the first cell of an AAL 5 frame. In this case, all subsequent cells belonging to the same AAL 5 frame are discarded except the last cell. This cell has to be kept so that the destination can determine the end of the AAL 5 frame. Early packet discard can be applied when the discarded cell is the first cell of an AAL 5 frame. In this case, all cells belonging to the same frame, including the last one, are discarded.

7.8 REACTIVE CONGESTION CONTROL

Reactive congestion control is based on a different philosophy to that used in preventive congestion control. In preventive congestion control we attempt to prevent congestion from occurring. This is done by first reserving bandwidth for a connection on each switch along the connection's path, and subsequently policing the amount of traffic transmitted on the connection. In reactive congestion control, at least in its ideal form, we let sources transmit without bandwidth reservation and policing, and we take action only when congestion occurs. The network is continuously monitored for congestion. If congestion begins to build up, a feedback message is sent back to each source requesting them to slow down or even stop. Subsequent feedback messages permit the sources to increase their transmission rates. Typically, congestion is measured by the occupancy level of critical buffers within an ATM switch, such as the output port buffers in a nonblocking switch with output buffering.

The Available Bit Rate (ABR) service, described below, is the only standardized ATM service category that uses a reactive congestion control scheme.

7.8.1 The Available Bit Rate (ABR) Service

This is a feedback-based mechanism, whereby the sending end device is allowed to transmit more during the time that there is a slack in the network. At connection set-up time, the sending end device requests a Minimum Cell Rate (MCR). It also specifies a maximum cell rate, which is its PCR. The network accepts the new connection if it can satisfy its requested MCR. We note that the MCR may also be zero. The transmission rate of the source may exceed its requested MCR, if the network has a slack capacity. When congestion begins to build up in the network, the sending end device is requested to decrease its transmission rate. However, its transmission rate will never drop below its MCR. The ABR service is not intended to support real-time applications.

It is expected that the sending end device is capable of increasing or decreasing its transmission rate according to the feedback messages it receives from the network. Also, it is expected that the sending end device that conforms to the feedback messages received by the network, will experience a low cell loss rate, and it will obtain a fair share of the available bandwidth within the network.

The control mechanism through which the network can inform the source to change its transmission rate is implemented using *Resource Management* (RM) cells. These are ATM cells whose Payload Type Indicator (PTI) is set to 110 (see Table 4.2). As shown in Figure 7.18, the transmitting end device generates *forward* RM cells which travel through the network to the receiver following the same path as its data cells. The receiver turns around these RM cells and transmits them back to the sending end device as *backward* RM cells. These backward RM cells follow the opposite path to the sending end device. (We recall that point-to-point connections are bidirectional.) ATM switches along the path of the connection may insert feedback control information in the RM cells, which is used by the sending end device to increase or decrease its transmission rate. Thus, a closed control loop is formed between the sending end device and its destination end device. This closed loop is used to regulate the transmission rate of the sending end device. A similar loop can be set up to regulate the transmission rate of the destination end device.

We note that feedback messages may not be effective when dealing with a link which has a long propagation delay. This is because the link may become a temporary storage capable of holding a large number of cells. As an example, let us consider a link which is 125 miles long connecting an end device to a switch, and let us assume that the end device transmits at 622 Mbps. This transmission speed translates to about 1462 cells per ms. Since light propagates through a fiber link at approximately 125 miles per ms, a maximum of 1462 cells may be in process of being propagated along the link. Let us now assume that at time t the switch sends a message to the end device requesting it to stop transmitting. Then, by the time the end device receives the message, the switch is likely to receive a maximum of 2×1462 cells. Of these cells, a maximum of 1462 cells can be already in flight at time t, and another maximum of 1462 cells can be transmitted

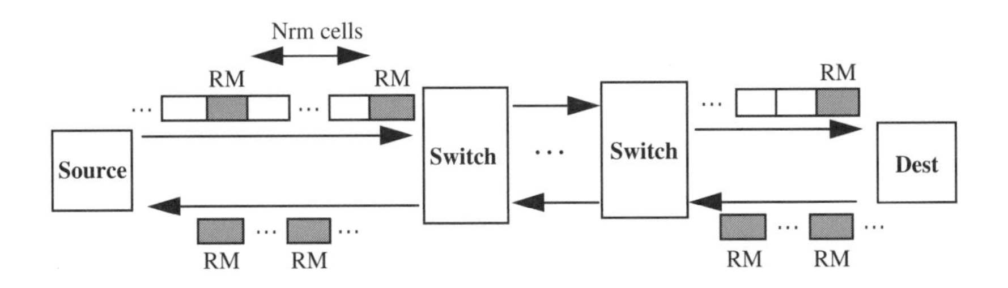

Figure 7.18 The ABR mechanism.

by the time the end device receives the message. To account for large propagation delays, manufacturers have introduced large buffers in their switch architectures. In addition, several feedback loops can be set-up, as shown in Figure 7.19, aiming at reducing the length of the control loop.

The ABR service does not include a formal conformance definition. However, verification that the source complies can be done using a dynamic GCRA, where the monitored transmission rate is modified based on the receipt of backwards RM cells.

RM Cell Structure

The RM cell's payload contains a number of different fields. Below, we describe some of these fields.

- *Message type field*: this is a one-byte field, and it contains the following 1-bit subfields:
  - *DIR*: this bit indicates the direction of the RM cell, i.e. whether it is a forward or a backward RM-cell.
  - *BN*: this bit indicates whether the RM cell is a *Backward Explicit Congestion Notification* (BECN) cell. As will be seen later, an ATM switch or the destination end device may independently generate a backward RM cell to notify the sending end device, instead of having to wait for an RM cell generated by the sending end device to come by. This RM cell has its BN bit set to 1. RM cells generated by the source, have their BN field set to zero.
  - *CI*: this congestion indication bit is used to by an ATM switch or the destination end device to indicate to the sending end device that congestion has occurred in the network.
  - *NI*: no increase indicator, used to prevent the sending end device from increasing its *Allowed Cell Rate* (ACR), which is its current transmission rate.
- *Explicit Rate (ER)*: this is a 2-byte field used to carry the explicit rate calculated by an ATM switch along the path. The ER is used to limit the sending end device's transmission rate. This field may be subsequently reduced by another ATM switch, if it calculates an ER which is lower than that indicated in the ER field of the RM cell.
- *Current Cell Rate (CCR)*: a 2-byte field used by the sending end device to indicate its ACR, i.e. its current transmission rate.
- *Minimum Cell Rate (MCR)*: this the minimum cell rate that the connection has requested and the network has agreed to guarantee.

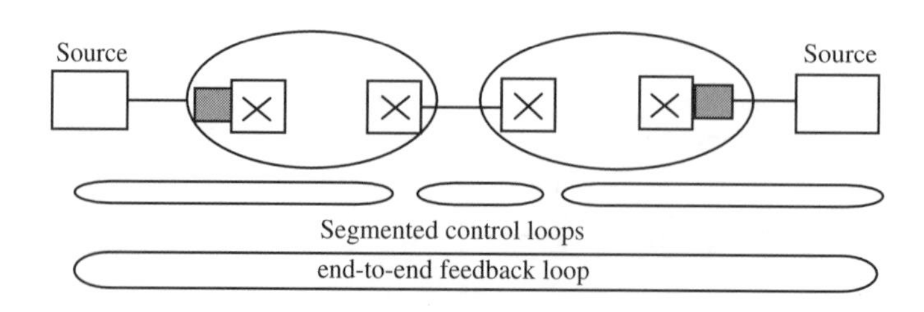

Figure 7.19 Feedback loops.

The ABR Mechanism

The source sends an RM cell every *Nrm* − 1 data cells. The defaulted value for *Nrm* is 32. The RM cells and data cells may traverse a number of switches before they reach their destination end device. The destination turns around the RM cells, which become backward RM cells, and transmits them back to the sending end device. Each switch writes information about its congestion status onto the backward RM cells, which eventually reach the sending end device. The feedback information send to the source depends on the mode of the ABR scheme. There are two modes, namely the *binary mode* and the *explicit rate mode*.

In the binary mode, the switch marks the EFCN bit in the header of the data cells to indicate pending congestion. (We recall that the EFCN bit is one of the three bits defined in the payload type indicator of the cell header.) The destination translates the EFCN information into bits such as the CI or NI, which are marked in the corresponding backward RM cell. Upon receipt of this, the source takes appropriate action. The action is (a) increase the transmission rate, (b) decrease the transmission rate, or (c) no change to the transmission rate. This mode is used to provide backward compatibility with ATM switches that conformed to earlier standards.

In the explicit rate mode, a switch computes a local fair share for the connection and marks the rate at which the source is allowed to transmit in the ER field of the backward RM cell. The switch does that only if the bandwidth it can offer to the connection is lower than what it is already marked in the backwards RM cell. The source, upon receipt of the backward RM cell, extracts the ER field and sets its transmission rate to the ER value. When detecting congestion, a switch can generate a backwards RM cell to convey the congestion status, without having to wait for a backwards RM cell to arrive.

Source Behavior

The source is responsible for inserting an RM cell every *Nrm* − 1 data cells. These RM cells are part of the source's Allowed Cell Rate (ACR). If the source does not have enough data cells to send, an RM cell is generated after a timer has expired and *Mrm* data cells have been transmitted. *Mrm* is fixed to 2. The data cells are sent with EFCN = 0.

The source adjusts its ACR according to the information received in an RM cell. ACR is greater than or equal to MCR and less than or equal to PCR. The ACR is adjusted as follows:

(a) If CI = 1, then the ACR is reduced by at least ACRxRDF, where RDF is a prespecified *rate decrease factor*. If the reduction results to a value below the MCR, then the ACR is set equal to the MCR.
(b) If the backward RM cell has both CI = 0 and NI = 0, then the ACR may be increased by no more than RIF × PCR, where RIF is a prespecified *rate increase factor*. The resulting ACR should not exceed the source's PCR.
(c) If the backward RM cell has NI = 1, then the ACR is not increased.

After ACR has been adjusted as above, it is set to at most the minimum of ACR as computed above and to the ER field, but no lower than MCR.

Destination Behavior

When a data cell is received, its EFCN is saved in the EFCN status of the connection. Upon receiving a forward RM cell, the destination turns around the cell and transmits it back to the source. The DIR bit is changed from forward to backward, BN = 0, and the fields CCR, MCR, ER, CI and NI in the RM cell remain unchanged, except in the following cases:

(a) If the saved EFCN status of the connection is set, then the destination sets CI = 1 in the RM cell, and resets the EFCN state.
(b) If the destination is experiencing internal congestion, it may reduce the ER to whatever rate it can support and set either CI = 1 or NI = 1.

The destination may also generate a new backward RM cell, with CI = 1 or NI = 1, DIR = 1, and BN = 1. This permits the destination to send feedback information to the source without having to wait for a source-generated RM cell to come by. The rate of these backwards RM cells is limited to 10 cells/s.

Switch Behavior

At least one of the following methods is implemented in a switch:

(a) *EFCN marking*: the switch may set the EFCN bit in the header of the data cells.
(b) *Relative rate marking*: the switch may set CI = 1 or NI = 1 in forward and/or backward RM cells.
(c) *Explicit rate marking*: the switch may reduce the ER field of forward and/or backward RM-cells.

The first two marking methods are part of the binary mode, whereas the third one is for the explicit rate mode. The term binary is used because the switch provides information of the type: congestion/no congestion.

A switch may generate backwards RM cells, to send feedback information to the source, without having to wait for a source-generated RM cell. The rate of these backwards RM cells is limited to 10 cells/s. Its fields are marked as follows: CI = 1 or NI = 1, BN = 1, DIR = 1.

A switch may also segment the ABR closed loop using a virtual source and destination. This can be useful in cases where the loop between the source and destination involves many hops, or long haul links with a large propagation delay. In such a case, the time it takes for the RM cells to return to the source may be significant. This may impact the time required for the source to react to an RM cell.

The calculation of the ER has to be done in such a way so that the available bandwidth in the switch has to be shared fairly among all the competing ABR connections. A number of different algorithms for calculating the ER have been proposed in the ATM Forum standards.

In the binary mode operation, the switch has to decide when to raise the alarm that congestion is pending. If we consider a non-blocking switch with output buffering, then if congestion occurs at an output port, the number of cells in its associated output buffer will increase dramatically. Typically, there are two thresholds associated with this buffer. A low threshold, T_{low}, and a high threshold, T_{high}. When the number of cells goes over

T_{high}, the switch can start marking the EFCN bit of the data cells or turn on the CI or NI bit in a forward or backward RM cell. As the sources begin to react to the feedback information, the number of cells in the buffer will go down below T_{high}. However, the switch continues marking until the number of cells in the buffer goes below T_{low}. At that moment, the switch stops the binary marking.

Simulation studies have shown that, in a binary feedback scheme as the one presented above, it is possible that some connections may receive more than their fair share of the bandwidth. Let us consider the case where source A is very close to a switch, and source B very far away. Then A will react to the feedback information from the switch much faster than B. For instance, if congestion occurs in the switch, it will decrease its transmission rate quicker than B. By the same token, when the congestion is lifted, it will increase its transmission faster than B. As a result, source A may put through more traffic than B.

PROBLEMS

1. Consider a 64 Kbps voice connection transmitted at constant bit rate (silence periods are also transmitted).
   (a) What is its PCR?
   (b) What is its SCR?
   (c) What is its average cell rate?
2. This is the continuation of problem 4, Chapter 5. On the average, a voice source is active (talkspurt) for 400 ms and silent for 600 ms. Let us assume that a voice call is transported over an ATM network via AAL 2. The voice is coded to 32 Kbps and silent periods are suppressed. We assume that the SSCS has a timer set to 5 ms. That is, each time the timer expires, it sends whatever data it has gathered to CPS as a CPS-packet. In problem 4, Chapter 5, you were asked to calculate the length of each CPS-packet, and the number of CPS-packets produced in each active period. Using this information, answer the following two questions:
   (a) What is the peak and average transmission bit rate of the voice source including the CPS-packet overhead?
   (b) What is the peak and average transmission rate of the voice source including the overheads due to the CPS-packet, the CPS-PDU and the ATM cell, assuming one CPS-packet per CPS-PDU?
3. Consider an on/off source where the off period is constant and equal to 0.5 ms. The MBS of the source is 24 cells. During the on period, the source transmits at the rate of 20 Mbps.
   (a) What is its PCR?
   (b) What is the maximum length of the on period in msec?
   (c) Assuming a 1 ms period, calculate its SCR.
4. Explain why the end-to-end cell transfer delay consists of a fixed part and a variable part. What is the fixed part equal to?
5. Explain why jitter is important to a delay-sensitive applications.
6. Consider an on/off source with a peak bit rate = 500 Kbps, and an average on period = 100 ms. The source will be transmitted over the output port of a non-blocking switch, which has a buffer $K = 500$ cells. Plot the average bit rate and the equivalent bandwidth of the source as a function of r, i.e. the fraction of time that the source is active. You should observe that the equivalent bandwidth tends to its peak bit rate when the source is bursty and to its average bit rate when the source is regular, i.e. not bursty. (Remember to express K in bits!)
7. Consider the virtual scheduling algorithm for policing the PCR. Assume that T (1/PCR) = 40 units of time and the CDVT = 60 units of time. The arrival times are: 0, 45, 90, 120, 125, 132, 140 and 220. Which of these arrivals will get tagged?
8. Repeat problem 7 using the continuous state leaky bucket algorithm.

APPENDIX: A SIMULATION MODEL OF AN ATM MULTIPLEXER—PART 2

This simulation project is the continuation of the simulation project entitled 'A simulation model of an ATM multiplexer—Part 1' described at the end of Chapter 6. In Part 1, you were asked to plot the cell loss rate as a function of the arrival rate which was controlled by the probability p that a slot contains a cell. You should have observed that the cell loss rate of an ATM multiplexer increased, as the arrival rate of each source feeding the multiplexer increased.

The objective of this project is to show that the cell loss rate of an ATM multiplexer increases as the burstiness of each source increases, while keeping the peak and average cell rate of each source constant. This will be demonstrated by extending task 2 of the previous simulation project to allow for bursty arrival sources, modelled by an Interrupted Bernoulli Process (IBP).

You can either write your own simulation program following the instructions given below, or use a simulation language.

Project Description

You will use the simulation model that you developed for task 2 in the previous simulation model. The structure of this simulation model will remain the same, except for the arrival processes. Specifically, in the previous simulation project, you assumed that the arrival process for each stream was Bernoulli. In this project, you will replace the arrival process for each stream by an IBP. All four arrival processes will be identical.

We will assume that time is slotted, and each slot is long enough so that a cell can completely arrive. As described in Section 7.1.3, an IBP is an on/off process defined over a slotted time axis. The on and off periods are geometrically distributed. That is, given that in slot i the process is in the on period, then in the next slot $i+1$ it will be still in the on period with probability p, or it will change to the off period with probability $1-p$. Likewise, if it is in the off period in slot i, it will stay in the off period in the next slot $i+1$ with probability q, or it will change to the on period with probability $1-q$. If in the next slot it will be in the on period, then a cell may arrive with probability a.

To simplify the traffic model, we will assume that each slot during the on period contains a cell. That is, $a = 1$. This traffic model can be completely defined if we know p and q. These probabilities can be obtained using the PCR, the average cell rate and the burstiness of the source as follows.

The burstiness of the source is defined by the squared coefficient of variation c^2. This quantity is defined as the variance of the inter-arrival time of cells divided by the mean inter-arrival time squared. For the above traffic model, it is given by the following expression:

$$c^2 = \frac{(1-p)(p+q)}{(2-p-q)^2} \qquad (1)$$

The larger its value, the burstier is the source. For instance, compressed voice has a c^2 of 19. To create extremely bursty sources, c^2 can be set as high as 200.

We also have that

$$\text{Average cell rate} = \text{PCR} \times \frac{\text{Average on period}}{\text{Average on period} + \text{Average off period}}$$

where

$$\text{Average on period} = \frac{1}{1-p}, \text{ and}$$

$$\text{Average off period} = \frac{1}{1-q}$$

Substituting these two expressions for the average on and off periods into the above expression for the average cell rate, we have:

$$\text{Average cell rate} = \text{PCR} \times \frac{1-q}{2-p-q} \tag{2}$$

We can estimate the parameters p and q of an IBP (assuming that during the on period cells arrive back-to-back), using expressions (1) and (2). For instance, let us consider a source with a peak bit rate of 1.544 Mbps, an average bit rate of 772 Kbps, and a c^2 of 20. Then, from (2) we have

$$\frac{1-q}{2-p-q} = 0.5$$

and from expression (1) we have:

$$\frac{(1-p)(p+q)}{(2-p-q)^2} = 20$$

After some calculations we can obtain that $p = q = 0.9756$.

Structure of the Simulation

The structure of the simulation will remain the same as in task 2. The generation of an arrival will be done according to an IBP, whose PCR, average cell rate and c^2, will be specified as input to the simulation. Given a PCR, average cell rate and c^2, you should first calculate p and q. An IBP can be simulated as follows.

Draw a pseudo-random number r, where $0 < r < 1$. If in current slot it is in the on period, then in the following slot it will remain in the on period if $r < p$. Else, it will shift to the off period. On the other hand, if it is in the off period, then in the next slot it will remain in the off period if $r < q$. Else, it will shift to the on period. If it will be in the on period in the next time slot, then we have an arrival.

Once a cell has arrived, draw a new pseudo-random number r, $0 < r < 1$, to decide which queue it will join. For that, follow the logic outlined in the previous simulation project.

Results

Calculate the cell loss probabilities per QoS queue and the total cell loss rate, for different values of c^2, and plot out your results. (If the curves are not smooth, increase the simulation run. This may happen as c^2 increases.)

APPENDIX: ESTIMATING THE ATM TRAFFIC PARAMETERS OF A VIDEO SOURCE

An MPEG video encoder generates frames which are transmitted over an ATM link. Write a program to simulate the traffic generated by the MPEG video encoder with a view to characterizing the resulting ATM traffic.

Problem Description

An MPEG video encoder generates frames with a group of pictures consisting of 16 frames. The first frame is an I-frame, and the subsequent 15 frames are P-frames. For a specific video clip, the number of bits X_n generated for the nth frame can be obtained as a function of the number of bits generated for the previous 16 frames, using the following auto-regressive expression:

$$
\begin{aligned}
X_n = {} & 0.412X_{n-1} + 0.12X_{n-2} + 0.11X_{n-3} + 0.07X_{n-4} + 0.06X_{n-5} + 0.05X_{n-6} \\
& + 0.001X_{n-7} + 0.032X_{n-8} - 0.001X_{n-9} + 0.001X_{n-10} - 0.032X_{n-11} \\
& - 0.002X_{n-12} - 0.05X_{n-13} - 0.041X_{n-14} - 0.1X_{n-15} + 0.37X_{n-16} + e_n
\end{aligned}
$$

where e_n is white noise and it follows the distribution $N(0, \sigma^2)$, with $\sigma^2 = 20\,000$ bits. The following initial values are used: $X_1 = 150\,000$ bits, and $X_i = 70\,000$ bits for $i = 2, 3, \ldots, 16$.

An I- or P- frame is generated every 30 ms. The information generated for each frame is transmitted over an ATM link using AAL1 unstructured PDUs. Assume that it takes zero time to pack the bits of a frame into AAL1 PDUs and subsequently into ATM cells. Also, assume that the ATM cells generated by a single frame are transmitted out back-to-back over a slotted link, with a slot equal to 3 μs.

Assume now that you are observing the ATM cells transmitted out on this slotted link. Due to the nature of the application, you will see that the ATM traffic behaves like an on/off model. You are required to measure the following parameters of this ATM traffic: average cell rate, sustained cell rate with $T = 900$ ms, MBS, average off period, and the squared coefficient of variation of the inter-arrival c^2.

Simulation Structure

The simulation program can be organized into three parts. In the first part, you generate the size of the next frame, and in the second part you collect statistics on the ATM cells generated by this frame. You will repeat these two parts until you have generated 5000 frames. Then, you will go to part 3, where you will calculate and print the final statistics.

Part 1

Use the above auto-regressive model to generate the size (in bits) of the next frame. For this, you will need to keep the size of the previous 16 frames. Start generating from frame number 17 using the initial values X_i, $i = 1, 2, \ldots, 16$, given above. In addition to calculating the weighted sum of the previous 16 frames, you will also need to generate an estimate for e_n. This is a random variate drawn from the distribution $N(0, \sigma^2)$, with $\sigma^2 = 20\,000$ bits. It can be generated using the following procedure:

1. Draw two random numbers r_1 and r_2, $0 < r_1, r_2 < 1$.
   Calculate $v = 2r_1 - 1$, $u = 2r_2 - 1$, and $w = v^2 + u^2$.
2. If $w > 1$ go back and repeat step 1, else, $x = v[(-2 \log_e w)/w]^{1/2}$.
3. Set $e_n = 141.42x$.

Part 2

A new frame is generated every 30 ms. Having generated the frame size, calculate how many ATM cells are required to carry this frame. Let X be the number of required ATM cells. These ATM cells are generated instantaneously, and they are transmitted out back-to-back, with a transmission time equal to 3 μs per cell. Calculate how many slots will be idle before the next frame arrives. Let the number of idle slots be Y. Update the following variables:

- frame_counter = frame_counter + 1
- total_simulation_time = total_simulation_time + 30
- total_cells_arrived = total_cells_arrived + X
- MBS = max{MBS, X}
- on_period = on_period + X
- off_period = off_period + Y

For the sustained rate, set-up a loop to calculate the total number of cells S arrived in 30 successive frames, i.e. in 900 ms. When 30 frames have been generated, compare this value against S, and save the largest of the two back in S. (Initially, set $S = 0$.)

To calculate c^2 you will need to keep all the inter-arrival times of the ATM cells. The inter-arrival is 1 between two cells that are transmitted back-to-back, and $Y + 1$ between the last cell of a frame and the first cell of the next frame. Maintain two variables, Sum and SqSum. For each inter-arrival time t, do the following:

Sum = Sum + t
SumSq = SumSq + t^2

If frame_counter < 5000, continue to generate frames, that is repeat parts 1 and 2. Otherwise go to part 3.

Part 3

Calculate and print out the required ATM traffic parameters:

- Average cell rate = total_cells_arrived/total_simulation_time
- SCR = $S/900$
- MBS
- average on period = on_period/frame_counter
- average off period = off_period/frame_counter
- c^2 = Var/MeanSq, where
  Var = [SumSq − (Sum^2/total_cells_arrived)]/(total_cells_arrived − 1)
  MeanSq = (Sum/total_cells_arived)^2

Part 3

Deployment of ATM

8

Transporting IP Traffic Over ATM

In this chapter, we present various solutions that have been proposed to carry IP traffic over ATM. We first present the ATM Forum's *LAN Emulation* (LE), a solution that enables existing LAN applications to run over an ATM network. Then, we describe IETF's schemes *classical IP and ARP over ATM* and *Next Hop Resolution Protocol* (NHRP), designed for carrying IP packets over ATM. The rest of the chapter is dedicated to the three techniques *IP switching*, *tag switching* and *Multi-Protocol Label Switching* (MPLS). IP switching utilizes the label swapping functionality of an ATM switch to transport IP packets in an efficient manner. IP switching had a short life span, but it inspired the development of tag switching, which has been standardized by IETF under the name of multi-protocol label switching. Tag switching, and also MPLS, have been primarily designed for IP networks, but they can also be used for ATM networks.

8.1 INTRODUCTION

In recent years, we have witnessed a tremendous growth in the number of hosts attached to the Internet. As the Internet traffic increases, the need to route IP packets faster increases as well. Several solutions have been proposed to switch IP traffic, ranging from gigabit routers to using the switching capability and functionality of ATM.

Since the early 1990s, there has been a quest for finding a solution to the problem of transporting connectionless traffic over ATM, which is inherently connection-oriented. The IETF and the ATM Forum have proposed several techniques for using the ATM network to transport IP packets, such as *LAN Emulation*, *classical IP and ARP over ATM*, and the *Next Hop Resolution Protocol* (NHRP). The development of these techniques was motivated by the desire to introduce ATM technology with as little disruption as possible to the existing IP model.

LAN emulation, as the name implies, emulates the characteristics and behavior of a LAN over an ATM network. It allows existing LAN applications to run over an ATM network without any modifications. LAN emulation was considered to be a good solution for providing faster connectivity to the desktop, since ATM could run at speeds such as OC-3, 100 Mbps TAXI, and 25 Mbps. These transmission speeds should be contrasted with the 10 Mpbs Ethernet that was available at that time, which due to software bottlenecks had an effective bandwidth of around 2 Mbps. LAN emulation was implemented and deployed successfully in the field. However, it never became the dominant technology in the LAN environment. This was due to the advent of the 100 Mbps Ethernet, which provided a high-speed connectivity to the desktop without having to change the underlying transport technology. The dominance of Ethernet in the LAN environment was further strengthened with the advent of Ethernet switches, and later on with the advent of gigabit Ethernet.

Classical IP and ARP over ATM was developed for a single IP subnet, that is, for a set of IP hosts that have the same IP network number and subnet mask. The members of the subnet communicate with each other directly over ATM, and they communicate with IP hosts outside their subnet via an IP router. Address resolution within the subnet is an important function of the protocol. This is necessitated by the fact that IP addresses are different to ATM addresses. Thus, there is a need to translate the IP address of a host to its corresponding ATM address, and vice versa. Finally, the Next Hop Resolution Protocol (NHRP), pronounced 'nerp', is an address resolution technique for resolving IP addresses with ATM addresses in a multiple subnet environment.

A different approach to switching IP packets, referred to as *IP switching*, was proposed by Ipsilon Networks (Ipsilon was later on purchased by Nokia). This technique is based on the label swapping functionality of an ATM switch. IP switching inspired the development of CISCO's *tag switching*, which was designed primarily for IP routers. Tag switching was proposed to circumvent the CPU-intensive table look-up in the forwarding routing table necessary to determine the next-hop router of an IP packet. It was also proposed as a means of introducing quality-of-service in the IP network. Tag switching served as the basis for a new protocol known as *Multi-Protocol Label Switching* (MPLS). This is an exciting new protocol that has been developed by IETF. Interestingly enough, since the introduction of tag switching, several CPU-efficient algorithms for carrying out look-ups in the forwarding routing table have been developed. The importance of MPLS, however, was by no means diminished, since it is regarded as a solution for introducing quality-of-service into the IP networks.

8.2 LAN EMULATION (LE)

LAN emulation is a service developed by the ATM Forum that enables existing LAN applications to run over an ATM network. This service emulates the characteristics and behavior of LANs. It provides connectionless service, broadcast and multicast service as supported by shared media LANs, it maintains the MAC address identity of each individual device attached to the LAN, and it allows existing LAN applications to work unchanged. LAN emulation provides interconnectivity over an ATM network among 802.3 Ethernet LANs, 802.5 token ring LANs, ATM stations, servers, and stations attached to LANs.

LAN applications and protocols run on top of the Logical Link Control (LLC) layer, described in Section 2.6. The LLC layer requires services from a MAC layer. If the LLC layer is kept the same, then the applications and protocols running on top of it do not have to change. In LAN emulation, the MAC layer is replaced by a LAN Emulation (LE) layer which provides MAC service to LLC and which runs on top of ATM, as shown in Figure 8.1. This solution permits various protocols such as IP, IPX, DECnet and Appletalk, to run in an emulated LAN.

LAN emulation allows an application to run on a computer which has an ATM interface and is directly connected to the ATM network, as shown in Figure 8.2. LAN emulation also allows a station attached to a LAN, such as Ethernet, to communicate with another station which is attached directly to an ATM network. This communication is enabled

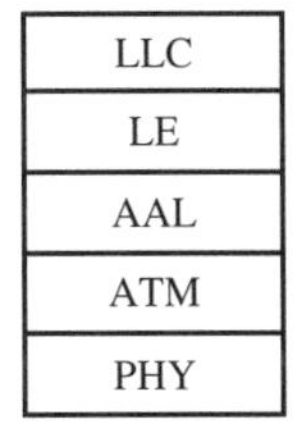

Figure 8.1 The protocol stack with the LE layer.

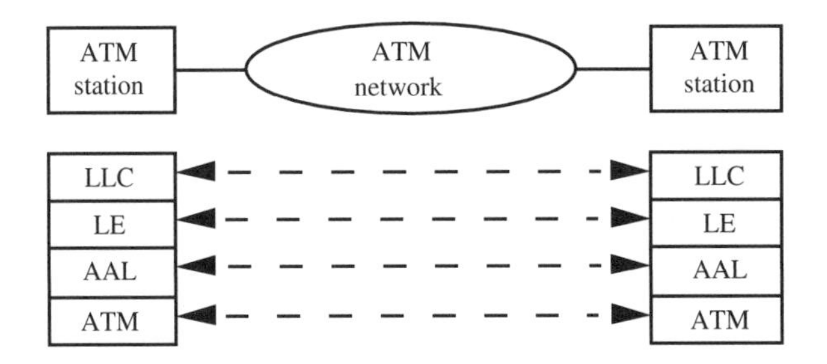

Figure 8.2 An application can run on an ATM station.

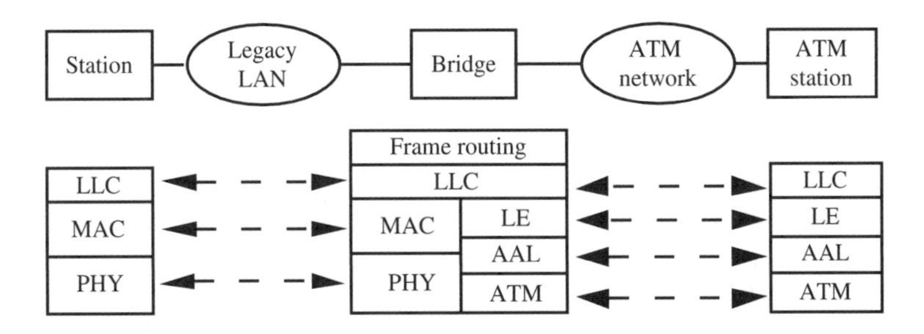

Figure 8.3 A station attached to a LAN can communicate with an ATM station.

through a bridge, as shown in Figure 8.3. LANs could be interconnected via an ATM network using LAN emulation, as shown in Figure 8.4.

An emulated LAN is a single segment LAN that can either be an Ethernet or a token ring. Membership in an emulated LAN is logical, whereas in a LAN membership is defined by who is physically attached to it.

A key component in LAN emulation is the *LAN emulation client* (LE client). The LE client resides at the end device, and it provides a MAC level emulated IEEE 802.3 Ethernet or 802.5 token ring interface to LLC. It performs functions emulating an IEEE 802.3 or 802.5 LAN, such as control functions data forwarding and address resolution.

The interaction between LE clients and the LE servers is done via the *LAN emulation User to Network Interface* (LUNI), shown in Figure 8.5. The following services are

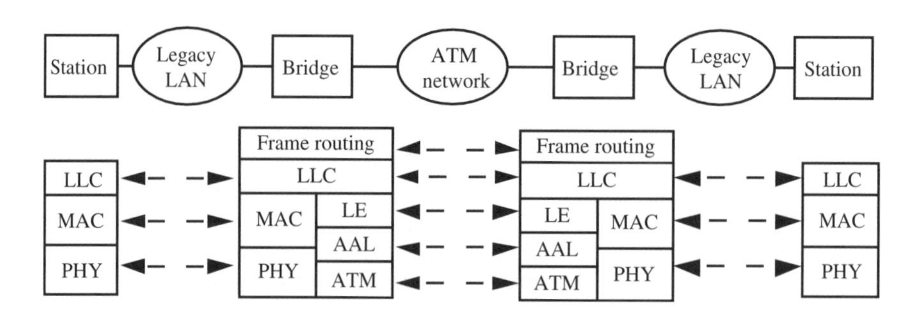

Figure 8.4 LANs could be interconnected via an ATM network.

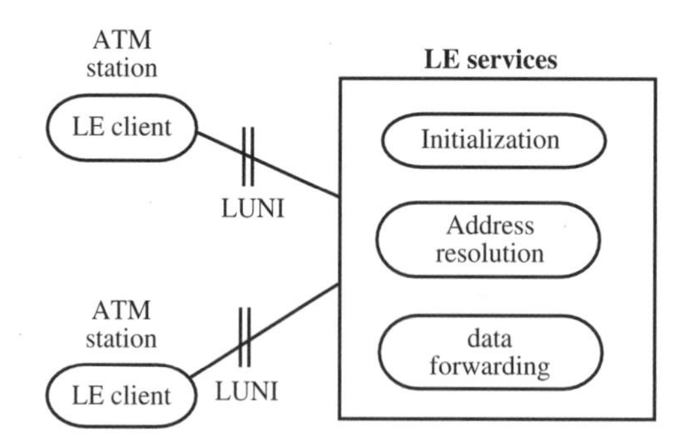

Figure 8.5 The LAN emulation User Network Interface (LUNI).

provided: initialization, registration, address resolution and data transfers. Initialization is used to obtain the ATM address of the LE services, and to join or leave a particular emulated LAN. Registration is used to inform the LE services of the list of individual MAC addresses that the LE client represents, and the list of multicast groups that the LE client belongs to. Address resolution is used to obtain the ATM address of an LE client with a known MAC address.

In addition to the LE client, the following LAN emulation components have also been defined:

- *LAN emulation server (LE server)*: the LE server provides an address resolution mechanism for resolving MAC addresses. Also, it performs functions such as registration of an LE client, forwarding address resolution requests, and managing LE client address registration information.
- *Multicast servers (MCS)*: one or more multicast server is used in an emulated LAN to provide the LE clients the connectionless data delivery characteristics found in a shared network. Its main task is to distribute data with a multicast address, to deliver initial unicast data before the destination ATM address has been discovered, and to distribute data with explorer source routing information.
- *Broadcast and Unknown Server (BUS)*: the BUS provides services to support broadcasting and multicasting, and initial unicast frames sent by an LE client before the

target ATM address has been resolved. This multicast server must always exist in an emulated LAN, and all the LE clients must join its distribution group. If there are no other multicast servers in the emulated LAN, the BUS handles all the multicast traffic.

An LE client has separate Virtual Circuit Connections (VCC) for control traffic and data traffic. The VCCs carry control or data traffic for only one emulated LAN, and they may be permanent or switched virtual circuits, or a mixture of both. A pictorial view of these VCCs is given in Figure 8.6.

Control VCCs link the LE client to the LE server and carry *LAN emulation address resolution* (LE-ARP) traffic and control frames. They are set-up during the initialization phase. The following types of control VCCs are used: *control direct VCC* and *control distribute VCC*. The control direct VCC is set-up by the LE client as a bidirectional point-to-point VCC to the LE server. It is maintained by both the LE client and the LE server as long as the LE client participates in the emulated LAN. The control distribute VCC is set-up by the LE server. It is an optional unidirectional control VCC to the LE clients, and it is used by the LE server to distribute control traffic. It may either be a point-to-multipoint VCC or a point-to-point VCC.

Data VCCs connect the LE clients to each other and to the multicast servers. The following types of data VCCs are used: *data direct VCC, multicast send VCC* and *multicast forward VCC*. A data direct VCC is a bidirectional, point-to-point VCC between two LE clients that want to exchange unicast data traffic. A multicast send VCC is a unidirectional, point-to-point VCC established by an LE client to an MCS or BUS. It is used by the LE client to send multicast data to an MCS or to the BUS. In the case of the BUS, it may also send initial unicast data. A multicast forward VCC is a unidirectional VCC from an MCS or the BUS to all the LE clients. It is used for distributing data to the LE clients. This may be a point-to-multipoint or a point-to-point VCC.

Address Resolution

When an LE client is presented with a frame for transmission to a destination LE client, it may or may not know the destination's ATM address. However, it knows the destination's

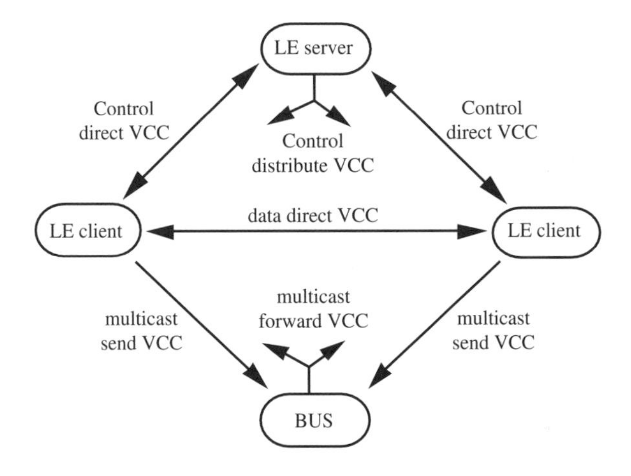

Figure 8.6 LAN emulation virtual circuit connections.

MAC address, since it will be passed on by the LLC. If it knows the ATM address, then it can establish a data direct VCC to the destination LE client, and then transmit the frame. If it does not know the ATM address, the LE client sends a LAN Emulation address resolution protocol (LE-ARP) request frame to the LE server over its control direct VCC. The LE-ARP request includes the source and destination MAC addresses, and the ATM address of the originating LE client. Since the LE server maintains a table of all MAC addresses and corresponding ATM addresses, it may be able to issue an LE-ARP reply to the requesting LE client. Alternatively, the LE server may forward the LE-ARP request to the appropriate LE client over the control distribute VCC or over one or more control direct VCCs. This will be for the case where the destination MAC address belongs to a workstation attached to a LAN on the other side of the bridge. All LE clients in the emulated LAN are required to accept this request. Each LE client checks the destination MAC address, and if it is his, it responds to the LE server over the control direct VCC with an LE-ARP reply. That reply is sent by the LE server over the control direct VCC to the originating LE client.

Alternatively, the requesting LE client can elect to transmit the frame to BUS through a multicast send VCC. The BUS then forwards the frame to the designated LE client.

8.3 CLASSICAL IP AND ARP OVER ATM

Classical IP and ARP over ATM is a technique standardized by IETF designed to support IP over ATM in a single *Logical IP Subnet* (LIS). A LIS is a group of IP hosts that have the same IP network address, say 192.43.0.0, and the same subnet mask, as shown in Figure 8.7(a). Now, let us assume that the LANs are replaced by three interconnected ATM switches, as shown in Figure 8.7(b). Each host can communicate directly with any other host in the subnetwork over an ATM connection. The traditional IP model remains unchanged, and the IP router is still used to connect to the outside of the subnet.

The term 'classical' in the name 'classical IP and ARP over ATM' is a reference to the use of ATM as a networking interface to the IP protocol stack operating in a LAN environment.

IP packets are encapsulated using IEEE 802.2 LLC/SNAP encapsulation. The protocol used in the payload, such as IP, ARP, Appletalk and IPX, is indicated in the LLC/SNAP header. An encapsulated packet becomes the payload of an AAL 5 frame. The Maximum

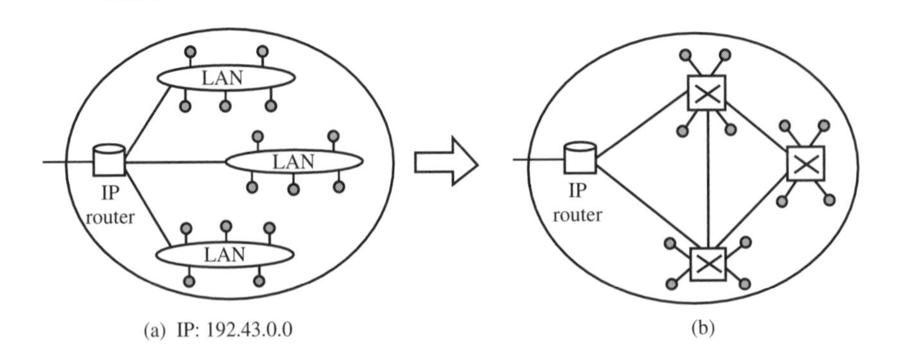

Figure 8.7 A Logical IP Subnet (LIS).

Transfer Unit (MTU) is fixed to 9180 bytes. Adding an 8-byte LLC/SNAP header gives a total of 9188 bytes, which is the default size for an AAL 5 frame.

8.3.1 ATMARP

Each member of the LIS is configured with an IP address and an ATM address. When communicating with another member in the same LIS over ATM, it is necessary to resolve the IP address of the destination host with its ATM address. IP addresses are resolved to ATM addresses using the *ATMARP* protocol within the LIS. This protocol is based on ARP (see Section 2.8), and it has been extended to operate over a nonbroadcast unicast ATM network. The *inverse ATMARP* (InATMARP) protocol is used to resolve an ATM address to an IP address. It is based on RARP (see Section 2.8), only it has been extended to support nonbroadcast unicast ATM networks.

The ATMARP protocol utilizes an ATMARP server which can run on an IP host or an IP router, and which must be located within the LIS. The LIS members are clients to the ATMARP server, and they are referred to as *ATMARP clients*. The ATMARP server maintains a table or a cache of IP and ATM address mappings. It learns about the IP and ATM addresses of ATMARP clients through a registration process, described below. At least one ATMARP server must be configured with each LIS. The following ATMARP messages have been defined:

- *ATMARP_request*: an ATMARP client sends an ATMARP request to the ATMARP server to obtain the ATM address of a destination ATMARP client. The message contains the client's IP and ATM addresses, and the IP address of the destination client.
- *ATMARP_reply*: this message is used by the ATMARP server to respond to an ATMARP_request with the requested ATM address. It contains the IP and ATM addresses of both the requesting and destination clients.
- *ATMARP_NAK*: negative response issued by the ATMARP server to an ATMARP_request.
- *InATMARP_request*: used to request the IP address of a destination. The message contains the sender's IP and ATM addresses and the destination's ATM address.
- *InATMARP_reply*: this is the response to an InATMARP_request with the destination's IP address. It contains the IP and ATM addresses of both the sender and the destination.

Registration

An ATMARP client must first register its IP and ATM addresses with the ATMARP server. This is done by invoking the ATMARP protocol as follows. Each ATMARP client is configured with the ATM address of the ATMARP server. After the client establishes a connection to the ATMARP server, it transmits an ATMARP_request on that connection. In the message, it provides its own IP and ATM addresses, and it requests the ATM address of itself by providing its own IP address as the destination IP address. The ATMARP server checks against duplicate entries in its table, time stamps the entry, and adds it to its table. It confirms the registration of the ATMARP client by sending an ATMARP_reply. If a client has more than one IP address within the LIS, then it has to register each IP address with the ATMARP server.

Entries in the table of the ATMARP server are valid for a minimum of 20 minutes. If an entry ages beyond 20 minutes without being updated (refreshed), then the entry is removed from the table. Each ATMARP client is responsible for updating its entry in the ATMARP server's table at least every 15 minutes. This is done by following the same procedure used to register with the ATMARP server, i.e. the ATMARP client sends an ATMARP_request to the ATMARP server with the destination IP address set to its own IP address. The ATMARP server updates the entry, and confirms it by responding with an ATMARP_reply.

Address Resolution

Let us assume that ATMARP client 1 wants to communicate with ATMARP client 2. We assume that both clients are in the same LIS. If there is already an established connection between the two clients, traffic can flow immediately. Otherwise, a connection can be set up if client 1 knows the ATM address of the destination client 2. If its destination ATM address is not known, client 1 sends an ATMARP_request to the ATMARP server. If the server has the requested address in its table, it returns an ATMARP_reply. Otherwise, it returns an ATMARP_NAK. Upon receipt of the ATMARP_reply, a connection is established and traffic starts flowing.

An ATMARP client creates an entry in its ATMARP table for every connection (PVCs or SVCs) that it creates. An entry is valid for a maximum of 15 minutes. When an entry has aged, the client must update it. If there is no open connection associated with the entry, then the entry is deleted. If the entry is associated with an open connection, then the client must update the entry prior to using the connection to transmit data. In the case of a PVC, the client transmits an InATMARP_request, and updates the entry on receipt of the InATMARP reply. In the case of an SVC, it transmits an ATMARP_request to the ATMARP server, and updates the entry on receipt of the ATMARP_reply.

An ATMARP client is also permitted to initiate the above procedure for updating an entry in the table, before the entry has aged.

8.3.2 IP Multicasting Over ATM

IP uses the class D address space to address packets to the members of a multicast group. Hosts and routers exchange messages using a group membership protocol called the *Internet Group Management Protocol* (IGMP). The routers use the results of this message exchange, along with a multicast routing protocol such as MOSPF, to build a delivery tree from the source subnetwork to all other subnetworks that have members in the multicast group.

In ATM, multicasting is implemented using a point-to-multipoint connection between a sending end device and multiple receiving end devices. In multicasting, the sender is known as the *root*, and the receivers are known as the *leaves*. An example of a point-to-multipoint connection is shown in Figure 8.8. The root is end device L1 and the leaves are end devices L2, L3, L4 and L5. As we can see, the connection is set up utilizing the multicasting feature of an ATM switch. Switch A receives cells from L1, and it transmits them out on two different links, namely on the link to switch C and on the link to switch D. Switch C receives cells from A, and it transmits them out to its end devices L2 and L3. Finally, switch D receives cells from A and transmits them to L5 and L6.

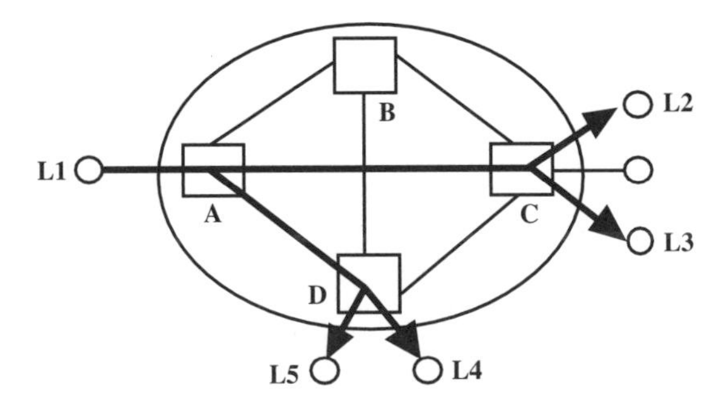

Figure 8.8 Multicasting in ATM.

A set of hosts that participate in an ATM-based multicast is referred to as a *cluster*. Two models exist to support an ATM-based multicast, namely, the *VC mesh* and the *multicast server* (MCS). In the VC mesh solution, a point-to-multipoint VC connection is set up for each host that wants to transmit multicast traffic. If all the hosts in a multicast cluster want to transmit and receive multicast traffic, then we will have a multicast tree for each host. That is, in the example shown in Figure 8.8, if the cluster consists of L1 to L5, then each end device will have a point-to-multipoint VC connection to the other end devices, with it being the root. In this case, each host is both the root and also a leaf in many point-to-multipoint VC connections. This criss-crossing of VC connections across the ATM network has given rise to the name of 'VC mesh'.

In the multicast server solution, each cluster member has a VC connection to a multicast server. The multicast server maintains a point-to-multipoint VC connection to all the cluster members. A host simply forwards its multicast traffic to the multicast server directly over its VC connection, which then multicasts it to the cluster members over its point-to-multipoint VC connection. An example of the multicast server solution is shown in Figure 8.9. When L1 wants to multicast traffic to the other cluster members L2 to L5, it sends it to the multicast server, which in turn multicasts it to all the cluster members. A side-effect of this scheme is that L1 will receive a copy of its own traffic back from MCS. An alternative solution is for the MCS to establish a point-to-point VC connection to each member of the cluster. This solution avoids the problem of a source receiving its

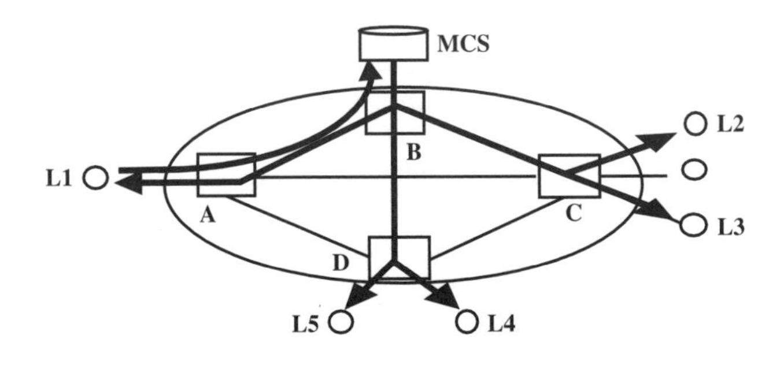

Figure 8.9 Multicasting in ATM.

own traffic. We note that the MCS solution is also used in LAN emulation, described in Section 8.2.

The VC mesh solution offers optimal performance, but it requires more VC connections than the MCS solution. In view of this, it does not scale up well. The MCS solution is easier to manage, but congestion bottlenecks may occur within the MCS. In IP and ARP over ATM, both the VC mesh and the multicast server models are implemented.

As we have seen in the previous section, where we discussed the unicast case, a mapping between the IP and ATM addresses of a host must be provided. This is done using the ATMARP server. A similar mapping is required in the multicasting case. This mapping is between an IP multicast group address and the ATM addresses of the members of the IP multicast group. This is done using the *Multicast Address Resolution Server* (MARS). MARS can be seen as the analog of the ATMARP server in multicasting. It supports a single cluster, and it can co-exist with the ATMARP server or an MCS on the same ATM switch. MARS does not participate in the actual multicasting of IP packets.

MARS maintains a table of IP group addresses, and for each IP group address it keeps all the ATM addresses of the cluster members that have registered with the specific multicast group. The table contains the following information:

$$\{\text{IP group address, ATM address 1, ATM address 2}, \ldots, \text{ATM address N}\}.$$

This table is known as the *host map*. The information in a host map is either configured manually or it is learned, as will be seen below.

The members of the cluster are clients to MARS. The MARS client protocol runs above the ATM adaptation layer and below the LLC layer, as shown in Figure 8.1. A MARS client can run on a IP host or an IP router. A MARS client that wants to join a particular IP multicast group establishes a VC connection to MARS. This connection is torn down if it is not used after a configurable period of time. The minimum suggested time is 1 minute, and the recommended default value is 20 minutes. The VC connection is used exclusively to send queries to MARS and receive replies from MARS.

The VC Mesh Scheme

MARS maintains a point-to-multipoint VC connection to all the members of a cluster, which is known as the *ClusterControlVC*, i.e. the cluster members are leaves of ClusterControlVC. This VC connection is used by MARS to distribute group membership information to the cluster members.

A MARS client that wishes to join a specific IP multicast group sends a MARS_JOIN message over the VC connection that it has established to MARS. A MARS_JOIN message contains the ATM address of the MARS client and the IP multicast address that it wants to join. When MARS receives a MARS_JOIN message, it adds the client as a leaf to the ClusterControlVC, and it then multicasts the MARS_JOIN message to all the cluster members associated with the particular multicast over the ClusterControlVC. The MARS client confirms that it has registered with the multicast group when it receives a copy of the MARS_JOIN message over the ClusterControlVC. Alternatively, MARS can confirm the registration of the client by sending back a copy of the message over the point-to-point VC connection between the client and MARS.

When a MARS client wants to leave a specific IP multicast group, it sends a MARS_LEAVE message to MARS. This message contains similar information as the

MARS_JOIN message, i.e. the client's ATM address and the IP multicast address that it wants to leave. When MARS receives the message, it removes the client from its ClusterControlVC, and then multicasts a copy of the message over the ClusterControlVC. MARS confirms that it has received and processed the message by sending back to the client a copy of the message over the point-to-point VC connection.

Now, let us see how this information that MARS collects about membership in the various IP multicast groups is used by a MARS client. When the IP layer in a host passes down an IP multicast packet, the host's MARS client ascertains whether a point-to-multipoint VC connection to the other cluster members that participate in the multicast exists. If it does not exist, it issues a MARS_REQUEST to MARS to request address resolution of the IP group address to which the multicast packet should be sent. MARS responds with a sequence of MARS_MULTI messages which contain the host map of the IP group address. A MARS_MULTI message carries as many ATM addresses as possible, but its length is limited to the Maximum Transfer Unit (MTU) of the underlying ATM connection. Therefore, depending upon the size of the host map, more than one MARS_MULTI message may be required. If MARS does not have a host map for the requested IP group address, it returns a MARS_NAK. Once the MARS client has the host map, it can create its own point-to-multipoint connection in order to multicast its IP packet. After transmitting the packet, the point-to-multipoint connection is kept open for any subsequent IP multicast packets.

As we have seen above, changes to the multicast membership are announced by MARS, so that the host can accordingly add or drop a leaf. The signaling procedures for setting up a point-to-multipoint VC connection and adding or dropping leaves are described in Chapter 10.

The Multicast Server (MCS) Scheme

In this scheme, it is the MCS that maintains a point-to-multipoint VC connection to the members of the cluster. A client simply forwards its traffic to the MCS, which in turn multicasts it over the point-to-multipoint VC connection. A MARS client registers with MARS following the same procedure as in the VC mesh case. However, the membership information is sent to the MCS, rather than to the cluster members. MARS announces membership information to the MCS over a connection known as the *ServerControlVC*. Since it is possible to have a number of MCSs, ServerControlVC is a point-to-multipoint connection, where the leaves are the MCSs. ServerControlVC is analogous to the ClusterControlVC.

An MCS must first register with MARS. For this, the MCS must have the ATM address of MARS which can be configured at start-up time of the MCS. After establishing a point-to-point VC connection to MARS, the MCS issues a MARS_MSERV message. When MARS receives this message, it adds the MCS to its ServerControlVC and returns a copy of the MARS_MSERV back to the MCS to confirm its registration. An MCS can drop from MARS by issuing a MARS_UNSERV message. MARS removes the MCS from its ServerControlVC, and returns a copy of the MARS_UNSERV message to confirm it.

During registration, no IP multicast groups are identified. An MCS can subsequently register with MARS to support one or more IP group addresses, again using a MARS_MSERV message. MARS confirms it by sending back a copy of the MARS_MSERV message. An MCS uses a MARS_UNSERV to specify to MARS that it does not want to support a specific IP group address. MARS confirms it by sending back a copy of the message. The confirmation messages that MARS sends back to the MCS are transmitted either over the ServerControlVC or the VC connection established between the MCS and MARS.

After an MCS registers with MARS to support an IP group address, it issues a MARS_REQUEST message to obtain the host map. MARS sends the information back in a sequence of MARS_MULTI messages. MARS sends a MARS_NAK if there is no host map. Subsequently, the MCS creates a point-to-multipoint VC connection to all the hosts that have registered with MARS for this particular multicast.

For each IP group address, MARS keeps two sets of mappings, namely the host map and the *server map*. The server map contains the information:

$$\{\text{IP group address, ATM address of MCS } 1, \ldots, \text{ATM address of MCS K}\}.$$

Typically, K is equal to 1, but it can be greater than one if more than one MCS is configured to serve an IP multicast group.

Now let us take a brief look at a host. A host follows the same procedure as in the VC mesh case in order to join or leave a multicast group. That is, it registers with MARS to participate in a specific multicast group using the MARS_JOIN, and it notifies MARS that it wants to leave a multicast group using the MARS_LEAVE message. Any changes in the multicast membership are reported to MARS, which subsequently announces them to the MCS server using MARS_SJOIN and MARS_SLEAVE messages. When a host sends a MARS_JOIN to MARS requesting to join a particular multicast group, MARS does not return the host map as in the VC mesh case, but it returns the server map. The client is lead to believe that the MCSs are the members of its multicast group, and it uses the server map to build a point-to-multipoint connection to its MCSs.

8.4 NEXT HOP RESOLUTION PROTOCOL (NHRP)

The IP model consists of networks which are interconnected by IP routers. Packets traveling from one network to another have to pass through an IP router. In classical IP and ARP over ATM, connectivity is limited to a single LIS. Traffic between two LISs has to pass through an IP router. In Figure 8.10, we give an example of how end devices A and B can communicate using classical IP and ARP over ATM. As can be seen, A is attached to LIS 1 and B to LIS 3. The two LISs communicate via IP routers 1 and 2. IP router 1 is attached to LISs 1 and 2, and IP router 2 is attached to LISs 2 and 3. Communication between A and B involves three separate VC connections. A communicates with IP router

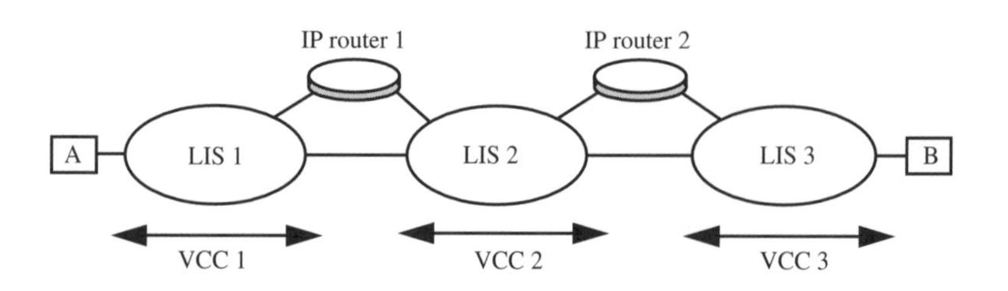

Figure 8.10 Classical IP and ARP over ATM.

1 via VC connection 1, IP routers 1 and 2 communicate via VC connection 2, and IP router 2 communicates with end device B via VC connection 3. Since VC connection 1 terminates at IP router 1, the router has to reassemble the original IP packets from the ATM cells, and then for each IP packet it has to do a table look-up in the forwarding routing table to determine the packet's next hop. The same applies to IP router 2. This, of course, introduces delays and additional work for the IP routers. Also, by terminating a connection at a router, the quality-of-service provided over this connection is terminated as well, since there are no quality-of-service guarantees within a router.

It is clear that some of the traffic would benefit if it were routed directly to the destination host without having to go through one or more IP routers. An example of this direct method is shown in Figure 8.11. A establishes a direct VC connection to B, thus bypassing the two IP routers along the way. This method can be useful for a source that requires a specific quality of service, whereas the classical IP and ARP over ATM method shown in Figure 8.10 can be used for sources which only require best effort.

The Next Hop Resolution Protocol (NHRP) is a technique proposed by IETF for resolving IP addresses to ATM addresses in a multiple subnet environment. It provides a host or an IP router with the ATM address of a destination IP address, so that a direct VC connection can be established. NHRP is not a routing protocol. It was originally designed as an extension to the classical IP and ARP over ATM, but it is not limited to IP networks.

NHRP is a master/slave protocol. The NHRP server is known as the *Next Hop Server* (NHS), and the NHRP client is known as the *Next Hop Client* (NHC). A NHRP server provides NHRP service for NHRP clients in a network which does not inherently support broadcasting or multicasting, and where the hosts and IP routers attached to the network can communicate with each other directly. Such a network is known as a *Non Broadcast Multiaccess Network* (NBMA). ATM, frame relay, X.25 and SMDS are examples of a NBMA.

A NHRP client must be attached to an ATM network, and it must know the ATM address of its NHRP server. NHRP clients can be serviced by one or more NHRP servers. A NHRP server can be located on a peer host, or on a default IP router attached to an ATM network. A NHRP server is configured with its own IP and ATM address, and a set of IP address prefixes that correspond to the domains of the NHRP clients that it serves. NHRP can work with any layer-3 internetworking protocol, such as IPX and Appletalk, over any NBMA network. The following NHRP messages have been defined:

- *NHRP next hop resolution request*: query sent from a NHRP client to a NHRP server requesting resolution of a destination IP address to an ATM address. It contains the IP and ATM addresses of the source, and the destination IP address.

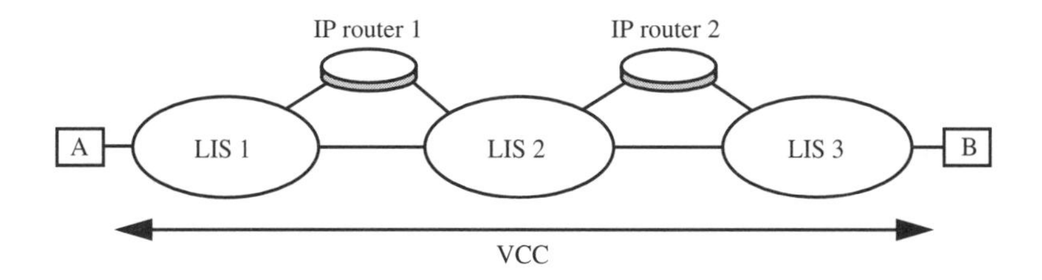

Figure 8.11 Direct connection.

- *NHRP next hop resolution reply*: response sent by a NHRP server to a query. It contains the IP and ATM addresses of the source, the IP and ATM addresses of the destination, and a NAK code.
- *NHRP registration request*: sent by a NHRP client requesting to register with the NHRP server. It contains the IP and ATM addresses of the source, and the IP address of the NHRP server.
- *NHRP registration reply*: response sent by a NHRP server to a registration request. It contains the IP and ATM addresses of the source, the IP address of the NHRP server, and a NAK code.
- *NHRP purge request*: used to invalidate cached information contained in a NHRP client or server. It contains the IP and ATM addresses of the source, and the IP address to be purged from the receiver's database.
- *NHRP purge reply*: sent in response to a NHRP purge request.
- *NHRP error indicated*: used to convey error information to the sender of a NHRP message. It contains the IP and ATM addresses of the source, and an error code.

NHRP Address Resolution

Let us consider a single NBMA network that contains a number of LISs, namely LIS1 and LIS 3. The two LISs are connected via LIS 2, as shown in Figure 8.12. End device A is attached to LIS 1, and it wants to establish a connection with end device B, which is attached to LIS 3. The two LISs are connected by IP routers 1 and 2, which also serve as NHRP servers for LIS 1 and LIS 3, respectively. The two IP routers are connected by a permanent virtual connection. The following steps will take place:

1. A sends a NHRP next hop resolution request message to NHRP server 1 with the information {A's ATM address, A's IP address, B's IP address}.
2. NHRP server 1 checks to see if it serves B. It also checks to see if it has an entry in its cache for B's IP address. If neither is true, it sends the NHRP next hop resolution request to the adjacent NHRP server 2.
3. NHRP server 2 receives the request from NHRP server 1, determines that it serves B's IP address, looks up its cache or table which contains the IP and ATM address of B, and returns the information to A over the path that the request came from. As the NHRP resolution reply goes back to A, NHRP server 1 may cache the information contained in the packet.
4. A sets up a direct VC connection to B, and data starts flowing.

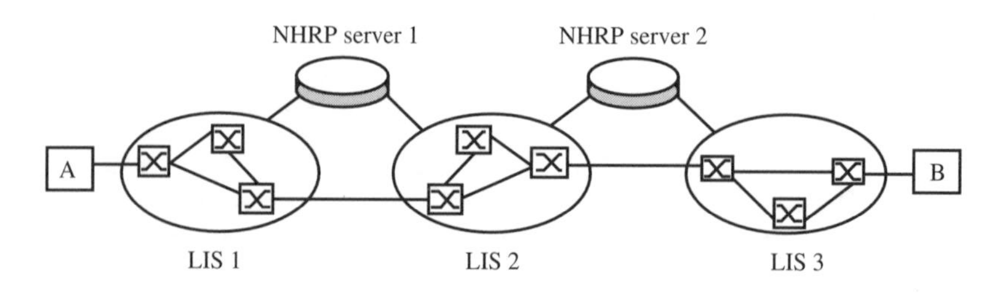

Figure 8.12 An example of address resolution.

The NHRP protocol is also used in the ATM Forum standard *Multi-Protocol Over ATM* (MPOA). This protocol integrates LAN emulation and NHRP, and it permits two devices attached to separate emulated LANs to set up an ATM connection and communicate directly.

8.5 IP SWITCHING

IP switching is an alternative technique to the schemes described so far in this chapter, and it was proposed by Ipsilon Networks. Similar schemes were also proposed by other companies. For instance, Toshiba proposed the *flow attribute notification protocol*, and CISCO proposed *tag switching*. These techniques are collectively known as *label switching* techniques. IP switching had a short life span, but it inspired the development of CISCO's tag switching, which has been standardized by IETF under the name of *Multi-Protocol Label Switching* (MPLS).

To understand the motivation behind this technique, let us consider the network shown in the Figure 8.13. It consists of an ATM network of five switches and of IP routers A, B and C. Each IP router is attached to an ATM switch. We assume that routers A and B are interconnected via a PVC using AAL 5. Likewise, routers B and C are also interconnected via a PVC using AAL 5.

Now, let us assume that IP router A has an IP packet to send to IP router C. From its forwarding table it determines that the next-hop router is B. The IP packet is encapsulated by AAL 5, and then it is segmented to an integer number of 48-byte blocks, each of which is carried in the payload of an ATM cell. The cells are forwarded to B over the PVC between A and B. At IP router B, the cells are reassembled into the original AAL 5 PDU, from which the original IP packet is extracted. B uses the IP address of the packet in its forwarding table and determines the next-hop IP router, which is C. The packet is subsequently encapsulated by AAL 5, and the resulting ATM cells are transmitted to C, where once more they get assembled back to the original IP packet.

IP switching can bypass the table look-up in the forwarding routing table of an IP router. We note that this table look-up was a time-consuming operation when IP switching was developed. This is no longer the case, since faster table look-up algorithms have been developed in the meantime.

In the above example, IP switching will forward the cells as they arrive at the ATM switch of IP router B directly to IP router C, without having to reassemble the cells into the original IP packet, and then carry out a forwarding routing decision at IP router B. This is not done for all IP packets, rather it is done for IP packets for which the router

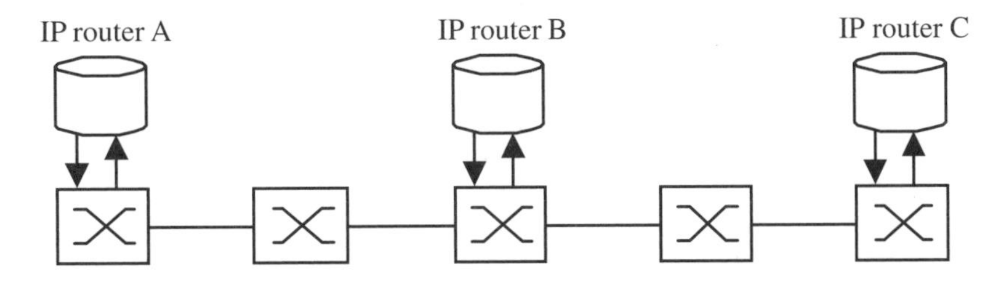

Figure 8.13 An example of three routers connected over ATM.

believes that they are part of an *IP flow*. This is a sequence of IP packets from a source IP address to a destination IP address. It can be identified by the pair ⟨source IP address, destination IP address⟩, or from a more detailed set of parameters, such as ⟨source IP address, source port number, destination IP address, destination port number⟩. This set of parameters used to identify an IP flow is referred to as the *IP flow id*. Using these parameters, the IP router can decide whether a particular IP packet it has received is an isolated packet, or whether it is the beginning of a sequence of IP packets. For instance, a DNS query will give rise to one or two IP packets, whereas an FTP is likely to give rise to a sequence of IP packets. Of course, there is no guarantee that an IP router will always guess correctly. For instance, it may determine that an IP packet is the beginning of an IP flow, but this packet may turn out to be just an isolated packet!

IP switching is implemented through *IP switches*. An IP switch is a general purpose computer which is attached to an ATM switch, and which runs the control and forwarding functions typically found in an IP router. In addition, it runs two protocols associated with IP switching, namely the *Ipsilon Flow Management Protocol* (IFMP) and the *General Switch Management Protocol* (GSMP). The functionality of these two protocols is explained below.

Two adjacent IP switches are connected over ATM via a default PVC. This connection is used to send control traffic, such as IP routing updates and IFMP messages. The same connection is also used to transmit ATM cells between the two IP switches. IP packets are first encapsulated using LLC/SNAP, and then they are transported over the ATM network using AAL 5. The LLC/SNAP encapsulation is merely used to indicate which layer 3 protocol is used.

To explain the basic principle of IP switching, let us consider three IP switches, namely A, B and C, connected as shown in Figure 8.14. A and B communicate via the default VCC 10/100, and B and C communicate via the default VCC 20/80. We assume that A sends traffic to B, which is then forwarded to C. All the cells transmitted from A to B have the virtual channel identifier 10/100. These cells are forwarded by the ATM switch to IP switch B. There, they get assembled back to the original AAL 5 PDUs, from which the IP packets are extracted. The IP switch looks up its forwarding table to identify the packet's next hop, and subsequently, the packet is encapsulated and transmitted out to C in a sequence of ATM cells. The VPI/VCI of these cells is 20/80.

Now, let us assume that IP switch B decides that a particular IP packet with an IP flow id x is the beginning of a new flow. B sends a REDIRECT message to A, requesting A to send the cells of the IP packets belonging to this flow x on a new VPI/VCI, say

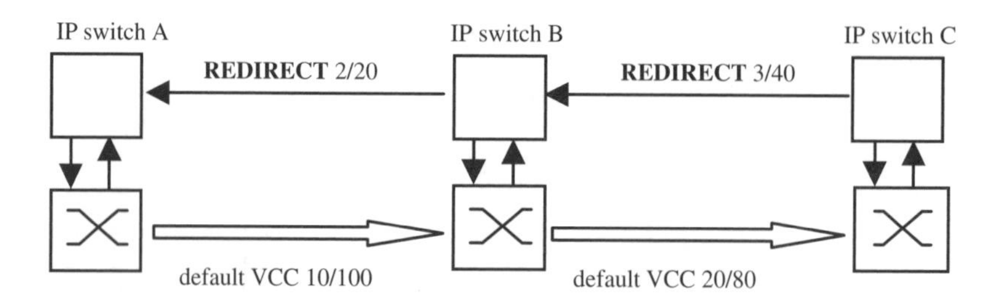

Figure 8.14 IP switching.

2/20. These cells still travel over the same link between A and B, but they are now distinguishable by their VPI/VCI. When these cells arrive at B's ATM switch, they are still forwarded to the IP switch, where as before, they are assembled into their original IP packets. The IP prefix is extracted, and B looks it up in its forwarding table to decide the next hop of the IP packet. The packet is then transmitted to C over the ATM network. So far, nothing has changed from the way IP switch B routes its packets.

Now, let us assume that C also decides to redirect IP flow x. It sends a REDIRECT message to B, asking B to send the ATM cells belonging to the IP packets with flow id x on a new VPI/VCI, say 3/40. These cells still travel over the same link from B to C, together with the rest of the traffic that B sends to C. However, they are distinguishable by their VPI/VCI.

At that moment, B recognizes that flow x has already been redirected from A to B on VPI/VCI 2/20. Therefore, it instructs its ATM switch to switch these cells directly through to the output port connecting to IP switch C, but with the VPI/VCI label set to 3/40. In view of this, all the cells belonging to the IP packets with flow id x simply cut through B's ATM switch without having to be forwarded to the IP switch.

If there is more than one IP switch between A and C, eventually all the IP switches will redirect the ATM cells associated with flow id x. As a result, these ATM cells will simply cut through all the ATM switches associated with these IP switches, without ever being forwarded to any of the intermediate IP switches.

We observe that IP switching is *data-driven*, i.e. a flow of IP packets has to be identified first before a cut through is set-up. Also, a redirect message is always sent by the IP switch which is downstream as far as the data flow is concerned. For instance, in the above example, it is IP switch C that request B to redirect the flow, and it is B that requests A to redirect the flow. As we will see in the next section, this is known in tag switching as *downstream allocation*.

Ipsilon Flow Management Protocol (IFMP)

This protocol runs between two IP switches, and it is used to communicate the redirection of an IP flow to a new VPI/VCI, otherwise known as the *label binding information*. It uses the default PVC, and it is a *soft-state* protocol, i.e. the state that it sets automatically times-out, unless it is refreshed. In view of this, the flow-binding information has a limited life once it is learned by an upstream IP switch, and it must be refreshed periodically as long as it is necessary. The messages that install flow states contain a lifetime field, which indicates for how long that state is to be considered valid. One advantage of this approach is that when the IP flow is finished, there is no need to cancel the binding upstream.

IFMP also provides an adjacency protocol, which can be used to identify immediate neighbors, and also make sure that a neighbor is alive.

The following five message are used by IFMP:

- *REDIRECT*: used to bind a VPI/VCI label to an IP flow.
- *RECLAIM*: used to release a VPI/VCI label for subsequent re-use.
- *RECLAIM ACK*: used to acknowledge that a RECLAIM message was received and processed.
- *LABEL RANGE*: used by an IP switch to communicate the acceptable range of VPI/VCI labels to its neighbors.
- *ERROR*: used for various error conditions.

General Switch Management Protocol (GSMP)

This protocol is used to control the ATM switch to which the IP switch is attached. The GSMP is a master/slave protocol, where the ATM switch is the slave and the IP switch is the master. The two systems communicate via an ATM link. The protocol allows the master to establish and release connections across the switch, add/delete leaves to point-to-multipoint connection, perform port management, and request statistics and configuration information. GSMP has an adjacency component and a connection management component.

8.6 TAG SWITCHING

This is a label switching technique proposed by CISCO, and has been standardized by IETF under the name of Multi-Protocol Label Switching (MPLS). In tag switching, a label is known as *tag*, which explains its name. Tag switching was developed primarily for IP networks, but it has also been applied to ATM networks.

An IP router implements both control and forwarding components. The control component consists of routing protocols, such as OSPF, BGP and PIM, used to construct routes and exchange routing information among IP routers. This information is used by the IP routers to construct the forwarding routing table, referred to as the *Forwarding Information Base* (FIB). The forwarding component consists of procedures that a router uses to make a forwarding decision on an IP packet. For instance, in unicast forwarding, the router uses the destination IP address to find an entry in the FIB, using the longest match algorithm. The result of this table look-up is an interface number, which is the output port connecting the router to the next-hop router, to which the IP packet should be sent.

A router forwards an IP packet according to its prefix (see Section 2.8.2). In a given router, the set of all addresses that have the same prefix, is referred to as the *forwarding equivalent class* (FEC), pronounced *fek*. IP packets belonging to the same FEC have the same output interface. In tag switching, it is a FEC that is associated with a tag. This tag is used to determine the output interface of an IP packet without having to look up its address in the FIB.

In IPv6, the tag can be carried in the flow label field. In IPv4, however, there is no space for such a tag in the IP header. If the IP network runs on top of an ATM network, the tag is carried in the VPI/VCI field of an ATM cell. If it is running over frame relay, the tag is carried in the DLCI field. For Ethernet, token ring and point-to-point connections running a link layer protocol such as PPP, the tag is carried in a special *shim* tag header, which is inserted between the LLC header and the IP header, as shown in Figure 8.15.

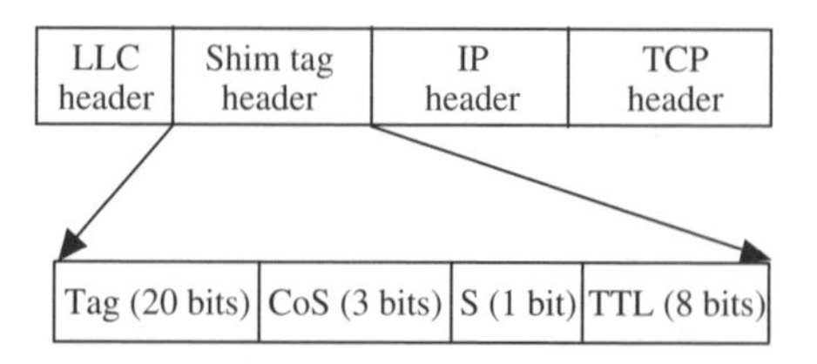

Figure 8.15 The shim tag header.

The first field of the shim tag header is a 20-bit field used to carry the tag. The second field is a 3-bit field used for the *Class-of-Service* (CoS) indication. This field is used to indicate the priority of the IP packet. The S field is used in conjunction with the tag stack. Finally, the *Time-to-Live* (TTL) field is similar to the TTL field in the IP header. The use of the CoS field and the tag stack will be explained later in this section.

We recall that the label switching mechanism in IP switching is data-driven, i.e. it is triggered by the arrival of an IP packet which is deemed to be the beginning of a sequence of packets. Tag switching, on the other hand, is *control-driven*, i.e. it is triggered when a router discovers a new FEC.

A tag switching network consists of *Tag Edge Routers* (TER) and *Tag Switching Routers* (TSR). A TER has the same functionality as a regular IP router, and in addition, it can bind tags to FECs. A TSR can only bind tags to FECs, and it cannot forward IP packets by carrying out a table look-up in the FIB.

To see how tag switching works, let us consider a network consisting of five TSRs, A, B, C, D and E, linked with point-to-point connections as shown in Figure 8.16. A, C and E are in fact tag edge routers. We assume that a new set of hosts with the prefix ⟨x.0.0.0, y.0.0.0⟩, where x.0.0.0 is the base network address and y.0.0.0 is the mask, is directly connected to E. The flow of IP packets with this prefix from A to E is via B and D. That is, A's next-hop router for this prefix is B, B's next-hop router is D, and D's next-hop router is E. Likewise, the flow of IP packets with the same prefix from C to E is via D. That is, C's next-hop router for this prefix is D, and D's next-hop router is E. The interfaces in Figure 8.16 show how these routers are interconnected. For instance, A is connected to B via if0, B is connected to A via if1, to C via if2 and to D via if0, and so on.

When a TSR identifies the FEC associated with this new prefix ⟨x.0.0.0, y.0.0.0⟩, it selects a tag from a pool of free tags and it makes an entry into a table known as the *Tag Forward Information Base* (TFIB). This table contains information regarding the incoming and outgoing tags associated with a FEC and the output interface, i.e. the FEC's next-hop router. The TSR also saves the tag in its FIB in the entry associated with the FEC.

The entry in the TFIB associated with this particular FEC for each TSR is shown in Table 8.1. (For presentation purposes, we have listed all the entries together in a single table.) We see that B has selected an incoming tag equal to 62, D has selected 15, and E has selected 60. A and C have not selected an incoming tag for this FEC, since they are tag edge routers, and they do not expect to receive tagged IP packets. The remaining

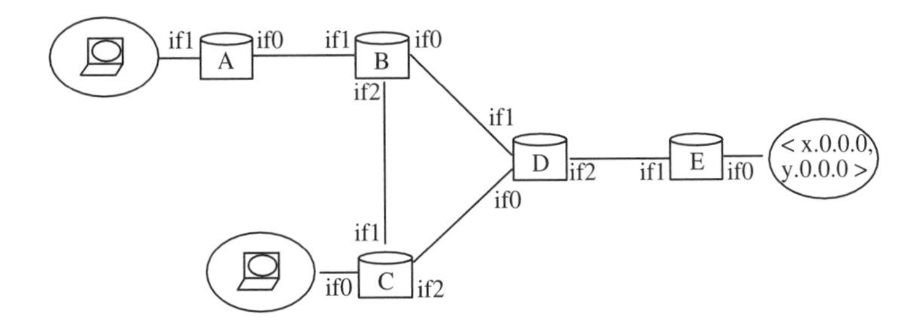

Figure 8.16 An example of tag switching.

Table 8.1 FEC entry in each TFIB.

TSR	Incoming tag	Outgoing tag	Next hop	Outgoing interface
A	—	—	TSR B	if0
B	62	—	TSR D	if0
C	—	—	TSR D	if2
D	15	—	TSR E	if2
E	60	—	TSR E	if0

information in each entry gives the next-hop router and the output interface for the FEC. For instance, for this FEC the next-hop router for A is B, and it is through if0.

An incoming tag is the tag that a TSR expects to find in all the incoming IP packets that belong to a FEC. For instance, in the above example, TSR B expects all the incoming IP packets belonging to the FEC associated with the prefix ⟨x.0.0.0, y.0.0.0⟩ to be tagged with the value 62. The tagging of these packets has to be done by the routers which are upstream of B, i.e. they are upstream in relation to the flow of IP packets associated with this FEC. In this example, the only router that is upstream of B is A. In the case of D, both B and C are upstream routers.

For a TSR to receive incoming IP packets tagged with the value that it has selected, the TSR has to notify its neighbors about its tag selection for a particular FEC. In the above example, TSR B sends its information to A, D and C. A recognizes that it is upstream from B, and it uses the information to update the entry for this FEC in its TFIB. D and C are not upstream from B as far as this FEC is concerned, and they do not use this information in their TFIBs. However, they store it for future use. It is possible, for instance, that due to failure of the link between C and D, B becomes the next-hop router for this FEC. In this case, C will use the tag advertised by B to update the entry in its TFIB.

D sends its information to B, C and E. Since B and C are both upstream of D, they use this information to update the entries in their TFIB. Finally, E sends its information to D, which uses it to update its entry in its TFIB. As a result, each entry in the TFIB of each TSR will be modified as shown in Table 8.2.

Table 8.2 FEC entry in each TFIB with tag binding information.

TFIB	Incoming tag	Outgoing tag	Next hop	Outgoing interface
A	—	62	TSR B	if0
B	62	15	TSR D	if0
C	—	15	TSR D	if2
D	15	60	TSR E	if2
E	60	—	TSR E	if0

We note that at E, there is no next-hop router for the FEC associated with the prefix ⟨x.0.0.0, y.0.0.0⟩. The IP packets associated with this FEC are forwarded to a local destination over if0.

Now, once the tags have been distributed and the entries have been updated in the TFIBs, the forwarding of an IP packet belonging to the FEC associated with the prefix ⟨x.0.0.0, y.0.0.0⟩ is done using solely the tags. Let us assume that A receives an IP packet from one of its local hosts with a prefix ⟨x.0.0.0, y.0.0.0⟩. A identifies that the packet's IP address belongs to the FEC, and it looks up its TFIB to obtain the tag value and the outgoing interface. It creates a shim tag header, sets the tag value to 62, and forwards it to the outgoing interface if0. When the IP packet arrives at TSR B, its tag is extracted and looked up in B's TFIB. The old tag is replaced by the new one, which is 15, and the IP packet is forwarded to interface if0. TSR D follows exactly the same procedure. When it receives the IP packet from B, it replaces its incoming tag with the outgoing tag, which is 60, and forwards it to interface if2. Finally, E forwards the IP packet to its local destination. The same procedure applies for an IP packet with a prefix ⟨x.0.0.0, y.0.0.0⟩ that arrives at C.

In Figure 8.17, we show the tags allocated by the TSRs. The sequence of tags 62, 15, 60 can be seen as being analogous to the VPI/VCI values allocated on each hop in an ATM VC connection. This sequence of tags can be seen as forming a path, referred to as the *tag switched path*, that resembles a point-to-point VC connection. An ATM connection is associated with two end devices, whereas a tag switched path is associated with a FEC. Typically, there may be several tag switched paths associated with the same FEC which form a tree, as shown in Figure 8.17.

Tag switching eliminates the CPU-intensive table look-up in the FIB, necessary to determine the next-hop router of an IP packet. A table look-up in the TFIB is not as time-consuming, since a TFIB is considerably smaller than a FIB. Since the introduction of tag switching, however, several CPU-efficient algorithms for carrying out table look-ups in the FIB have been developed. This did not diminish the importance of tag switching, since it was seen as a means of introducing quality of service in the IP network. This is done by associating a priority for each IP packet that gets tagged by a Tag Edge Router (TER). The priority is carried in the Class of Service (CoS) field shown in Figure 8.15.

Tagged IP packets within a TSR are served according to their priority, as in the case of an ATM switch. We recall that in ATM networks, each VC connection is associated with a quality-of-service category. An ATM switch can determine the quality-of-service of an incoming cell from its VPI/VCI value, and accordingly, it can queue the cell into

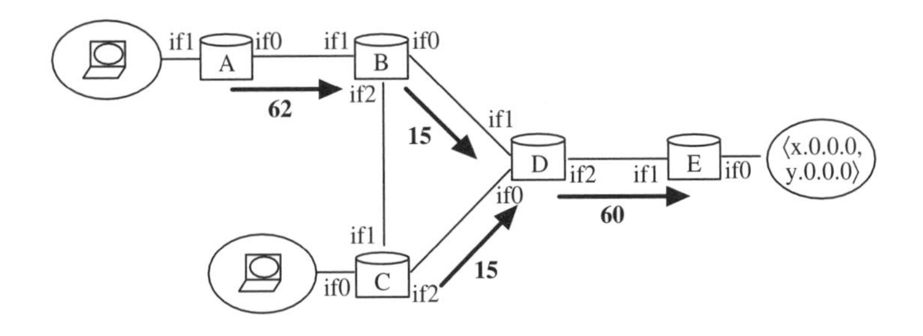

Figure 8.17 Tag switched paths.

the appropriate quality-of service queue. As we have seen in Section 6.7, an ATM switch maintains different quality-of-service queues for each output buffer. These queues are served using a scheduling algorithm, so that VC connections can be served according to their requested quality-of-service. A similar queueing structure can now be introduced into an IP router.

Another interesting feature of tag switching is that it can be used to create a dedicated route, known as an *explicit route*, between two IP routers. Explicit routing is described below.

Tag Allocation

In the example described above, a TSR allocates a tag for a FEC and saves this information in its TFIB as the incoming tag. It then advertises the binding between the incoming tag and the FEC to its neighboring TSRs. This information could be carried by either piggybacking it to a routing protocol, or using the *Tag Distribution Protocol* (TDP). When a TSR receives this information, it places the tag in the outgoing tag field of the entry in its TFIB that is associated with this FEC. In view of the fact that this information is generated by the TSR which is at the downstream end of the link, with respect to the flow of the IP packets, this type of tag allocation is known as *downstream tag allocation*. IP switching also uses downstream allocation. In addition to this scheme, tags can be allocated using *downstream tag allocation on demand*, and *upstream tag allocation*.

In downstream tag allocation on demand, each TSR allocates an incoming tag to a FEC, and creates an appropriate entry in its TFIB. However, it does not advertise its tag to its neighbors, as in the case of downstream allocation. Instead, an upstream TSR obtains the tag information by issuing a request via TPD.

Upstream tag allocation, as its name implies, works in the opposite direction to downstream allocation. When a TSR discovers a new FEC, it selects a tag and creates an appropriate entry in its TFIB. This tag is the outgoing tag for the FEC, rather than the incoming tag, as in the case of downstream tag allocation. The TSR advertises its tag for this FEC to its neighboring TSRs. A downstream TSR will populate the incoming tag field of its entry associated with the FEC with the advertised tag. If it is not a downstream TSR, it will simply store this information.

Tag Stack

The IP routing architecture consists of a collection of routing domains. Intra-domain routing (i.e. routing within a domain) is provided via an interior routing protocol, such as OSPF. Inter-domain routing (i.e. routing in-between domains) is provided by an exterior routing protocol, such as BGP. TSRs allow the decoupling of interior and exterior routing information, so that only the TSRs at the border of a domain are required to maintain routing information provided by the exterior routing protocol. TSRs within a domain maintain only routing information provided by the domain interior routing. (This is not currently the case in IP networks.)

To support this functionality, tag switching allows an IP packet to carry multiple tags organized as a stack. When a packet is forwarded from a border TSR of one domain to a border TSR of another domain, the tag stack contains a single tag. However, when the packet is forwarded within a domain, it contains two tags. The tag at the top of the stack is used for tag switching within the interior TSRs, so that the packet is forwarded to the egress border TSR. The tag in the lower level is used by the egress border TSR to forward the packet to the next border TSR.

Explicit Routing

As we have discussed above, a router makes a forwarding decision by using the IP address in its FIB to determine the next-hop router. Typically, each IP router calculates the next-hop router for a particular destination using the shortest path algorithm. Tag switching follows the same general approach, only it uses tags. This routing scheme is known as *destination-based* routing.

An alternative way of routing a packet is to use *source routing*. In this case, the originating (source) TSR selects the path to the destination TSR. Other TSRs on the path simply obey the source's routing instructions. Source routing can be used in an IP network to evenly distribute traffic among links, by moving some of the traffic from highly utilized links to less utilized links. Tag switching can be used to set-up such routes, which are known as *explicit routes*. The creation of an explicit route is done using RSVP.

Tag Switching Over ATM

In this scheme, the ATM user plane of an ATM switch remains intact. However, the ATM signaling protocols, such as Q.2931 and PNNI, are replaced by IP protocols such as OSPF, BGP, PIM and RSVP. An ATM switch which can also run tag switching is referred to as an ATM-TSR.

In tag switching over ATM, there is no shim tag header, as is the case when tag switching is implemented over Ethernet, token ring or a point-to-point connection. The tag is carried in the VCI field of the cell. If a tag stack is used, then up to two tags can be carried, one in the VCI field and the other in the VPI field. A predefined VC connection is used for exchanging tag binding information.

Downstream allocation on demand is used. That is, when an ATM-TSR identifies a new FEC it selects an incoming tag, but it does not advertised it to its neighbors. An upstream ATM-TSR obtains the tag bound to the FEC by issuing a request. Eventually, the connection will be set-up from an ATM-TSR that acts as an edge TSR to the edge TSR that serves the hosts associated with the FEC.

An interesting problem that arises in tag switching over ATM is *VC merging*. This problem arises in destination-based routing when two ATM-TSRs are both connected to the same downstream ATM-TSR. Let us consider the four ATM-TSRs A, B, C and D, shown in Figure 8.18, and let us assume that the flow of IP packets for a specific FEC is from A to C and then to D, and from B to C and then to D. We assume that D is the edge TSR that serves the hosts associated with the FEC. The allocated tags are shown in Figure 8.18 in bold. Now, let us see what happens when A has an IP packet, call it packet 1, to transmit that belongs to this FEC. This packet will be encapsulated by AAL 5, and then it will be segmented into an integer number of 48-byte blocks. Each block will then be carried in the payload of an ATM cell tagged with the value 15. The cells will be transmitted to C, where they will have their tag changed to 20, and then they will be forwarded to the buffer of the output port that connects to D. In this buffer, it is possible that these cells will get interleaved with the cells belonging to an IP packet, call

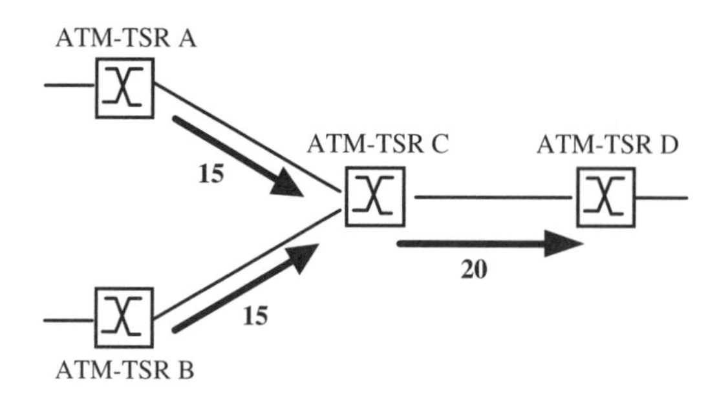

Figure 8.18 VC merging.

it packet 2, associated with same FEC and transmitted from B. That is, as the cells are queued-up into the buffer, cells belonging to packet 1 may find themselves in-between two successive cells belonging to packet 2. Since all these cells will be send to D with the tag of 20, D will not be able to identify which of these cells belong to packet 1 or packet 2, and consequently, it will be not be able to reconstruct the original AAL 5 PDUs.

A simple solution to this problem is to first collect all the cells belonging to the same IP packet in the buffer of the output port of C. Once all the cells have arrived, then they can be transmitted out back-to-back to D. To do this, it will be necessary for the switch to be able to identify the beginning cell and last cell of an AAL 5 PDU. The mechanism to do this may be in place if the early packet discard and partial packet discard policies have been implemented in the switch (see Section 7.7.2). An alternative solution is to use multiple tags, so that the path from A to D is associated with a different set of tags than the path from B to D.

8.7 MULTI-PROTOCOL LABEL SWITCHING (MPLS)

MPLS is an IETF standard. It is based on tag switching. The original intention was to be used in conjunction with different networking protocols, such as IPv4, IPv6, IPX and AppleTalk. However, MPLS has been developed exclusively for IP networks, which makes the description of the protocol as being a 'multi-protocol' more general than it actually is.

Basic Features of the MPLS Architecture

The main architecture of MPLS is the same as tag switching. A tag in MPLS is called a *label*, a Tag Switching Router (TSR) is called *Label Switching Router* (LSR), a Tag Edge Router (TER) is called a *Label Edge Router* (LER), and an ATM TSR is called ATM-LSR. Finally, a tag switched path is referred to as a *Label Switched Path* (LSP).

An LSR can only perform label-based functions. An LER terminates and originates LSPs, and it performs both conventional IP router functions and label-based functions. On the ingress to an MPLS domain, an LER accepts unlabeled IP packets and creates an initial MPLS label stack consisting of one or more shim label headers. On the egress of an LSP, an LER terminates and forwards the IP packet based on their IP addresses. A hybrid LSR originates and terminates some LSPs, while at the same time it acts as an LSR for other LSPs.

Both destination-based routing (called *hop-by-hop* in MPLS) and explicit routing is allowed. Explicit routing is referred to as *constraint routing*, because the route is picked up using a constraint other than the number of hops, such as minimize congestion along the path. A constraint route may not necessarily be the shortest path.

Label allocation in MPLS is control-driven, i.e. it is triggered when a router discovers a new FEC. (Label allocation can also be data-driven, as in IP switching.) The allocation of labels can be done using downstream allocation or downstream allocation on demand. As in tag switching, the label is carried in a *shim label header* if the IP network runs over Ethernet, token ring or a point-to-point protocol such as PPP. The location and structure of the shim label header is the same as the shim tag header shown in Figure 8.15. Multiple shim label headers can be stacked together in a *label stack*, as shown in Figure 8.19. All shim label headers in a label stack have S = 0, except the one at the bottom, which has S = 1.

When MPLS runs on top of ATM, the label is carried in the VCI/VPI field. The label stack is carried as part of the IP packet, where the block of information passed on to AAL 5 consists of the IP packet and the label stack placed in front of the IP header. The top label in the stack is the one used in the VCI/VPI field of the cell header. The remaining labels can be used in cases where the MPLS domain is extended past an ATM network into a non-ATM network which requires the use of the shim label header.

The distribution of labels can be done by piggy-backing them on control protocols such as BGP, PIM and RSVP. In addition, a new protocol for the distribution of labels, the *Label Distribution Protocol* (LDP), has been developed by IETF.

The Label Distribution Protocol (LDP)

For reliability purposes, the LDP protocol runs over TCP. Two LSRs which use LDP to exchange label-binding information are known as *LDP peers*. For two LDP peers to be able to exchange information, they have to establish an *LDP session* between them.

There are four categories of LDP messages: *discovery*, *session*, *advertisement* and *notification* messages. Discovery messages provide a mechanism whereby LSRs indicate their presence in a network by sending hello messages periodically. Session messages are used to establish, maintain and terminate an LDP session between LDP peers. Advertisement messages are used to create, change and delete label mappings for FECs. Finally, notification messages are used to provide advisory information and signal error information.

All LDP messages have a common structure that uses the *Type-Length-Value* (TLV) encoding. The type specifies how the value field is to be interpreted, the length gives the

Label (20 bits)	CoS (3 bits)	S (1 bit)	TTL (8 bits)
Label (20 bits)	CoS (3 bits)	S (1 bit)	TTL (8 bits)
⋮			
Label (20 bits)	CoS (3 bits)	S (1 bit)	TTL (8 bits)

Figure 8.19 The label stack.

length of the value, and the value field contains the actual information. The value field may itself contain one or more TLVs, i.e. TLVs may be nested.

For constraint routing, the label-binding information is distributed using RSVP or CR-LDP, an extension of LDP. CR-LDP can be used to trigger and control the establishment of an LSP between two LERs. *Strict* routing or *loose* routing can be used: in strict routing, all the LSRs through which the LSP must pass are indicated; in loose routing, some LSPs are indicated, and the exact path between two such LSPs is determined using conventional routing based on IP addresses.

PROBLEMS

1. What was ATM Forum's motivation for creating LAN emulation?
2. Identify two problems that LAN emulation has to resolve to provide LAN services over ATM.
3. Explain the function of each of the control VCCs used in LAN emulation.
4. Explain the function of ATMARP.
5. Explain the function of MARS.
6. In IP and ARP over ATM, describe the sequence of actions that take place when a new leaf is added to a multicast group, assuming an MCS-based architecture.
7. What problem does NHRP address?
8. What is an IP flow?
9. Explain why IP switching is data-driven.
10. Explain the differences between downstream and upstream allocation.
11. In MPLS, what is the label stack used for? Give an example.
12. Explain the difference between destination-based, or hop-by-hop, routing and explicit routing.

9

ADSL-based Access Networks

In recent years, a number of different technologies have been developed with a view to providing high-speed access to residential users and small offices. Specifically, a family of technologies known as xDSL has been developed to provide access to the Internet over the telephone line, in addition to basic telephone services. Cable modems have also been developed to provide access to the Internet over the TV cable. In addition, wireless access technologies, such as *Local Multipoint Distribution Services* (LMDS), make use of new wireless spectrums to provide high-speed access. Recently, *ATM Passive Optical Networks* (APON) have emerged as an alternative to providing high-speed access over fiber.

In this chapter, we first review these access technologies, and then present in detail the *Asynchronous Digital Subscriber Line* (ADSL) technology. We also discuss schemes for accessing *Network Service Providers* (NSP) over ADSL.

9.1 INTRODUCTION

Cable TV has been widely deployed in the US. Typically, each home is wired with a telephone line and a TV cable. The telephone line is a twisted pair that terminates to the nearest telephone switch, referred to as the *Central Office* (CO). The TV cable is a coax cable that connects the home to the cable distribution network. The telephone and cable networks are two separate systems: one receives TV channels over the TV cable, and uses the telephone for telephone services, and also to connect to the Internet over a modem. Due to the enormous business opportunities, both cable and telephone operators are interested in providing a complete set of services over the same wire. Currently, telephone operators provide access to the Internet over the twisted pair, in addition to basic telephone services. Video on demand will also be provided over the twisted pair in the future. Cable operators provide access to the Internet over their coax TV cable distribution system in addition to the distribution of TV channels. It is expected that telephone services (voice over IP) will also be provided over the cable. The battle between the telephone and the cable operators for dominance in the access network area will only intensify!

A cable plant architecture consists of the headend, fiber trunks extending from the headend, and coaxial cables. The headend is the source of the TV channels that are distributed over the cable plant. The homes served by the cable plant are divided into small neighborhoods of about 500 homes, and each neighborhood is connected to the headend via a dedicated fiber. Each fiber extends from the headend and terminates at an *Optical Network Unit* (ONU). From the ONU, a number of coaxial cables fan out into the neighborhood, each serving a number of homes, as shown in Figure 9.1. Due to the

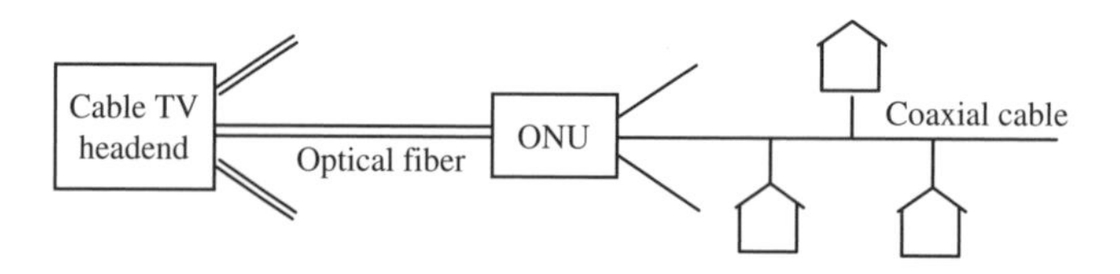

Figure 9.1 The HFC architecture.

combination of fiber optics and coaxial cables, this architecture is known as the *Hybrid Fiber Coaxial* (HFC) architecture. Access to the Internet is provided over an HFC plant using cable modems. These modems convert data packets to analog signals, which are transmitted over an analog path in the same way as analog video. Due to the enormous bandwidth of cable TV networks, access to the Internet can be achieved at speeds of 6 Mbps or higher. Moreover, existing cable TV systems have upstream channels built in for interactive services. These upstream channels can be used to provide a path to the IP network to which the headend is connected. The transmission of high-speed data over cable systems is specified in the *Data-Over-Cable Service Interim Specification* (DOCSIS), Cable-based access networks have been designed to transport IP traffic.

ADSL is one of the access technologies that can be used to convert the telephone line into a high-speed digital link. It is part of a family of technologies called the *x-type Digital Subscriber Line* (xDSL), where x stands for one of several letters of the alphabet, and it indicates a different transmission technique. Examples of the xDSL family technologies are: *Asymmetric DSL* (ADSL), *High data rate DSL* (HDSL), *Symmetric DSL* (SDSL), *ISDN DSL* (IDSL), and *Very high data rate DSL* (VDSL). Some of the xDSL technologies use analog signaling methods to transport analog or digital information over the twisted pair, while others use true digital signaling to transport digital information. A list of specifications for the xDSL family technologies is given in Table 9.1. In access networks, *downstream* means from the network to the user, and *upstream* means from the user to the network.

VDSL, as its name implies, achieves very high data rates over the twisted pair. However, the distance over which such rates can be transported is limited. Currently, it can achieve a downstream data rate of 52 Mbps and an upstream data rate of 6.4 Mbps over a distance of up to 1000 feet. For the same distance, it can also provide symmetric rates of 26 Mbps downstream and 26 Mbps upstream. The longest distance it can be transported is currently

Table 9.1 xDSL specifications.

Name	Downstream/upstream rate	Use
ADSL	8.128 Mbps/ 800 Kbps	Data
HDSL	1.544 /2.048 Mbps	T1/E1 replacement
SDSL	768 Kbps	Fractional T1/data
ISDL	128 Kbps	Data
VDSL	52 Mbps/6 Mbps	Video/data

5000 feet, for which it can achieve 13 Mbps downstream and 1.6 Mbps upstream. VDSL can be used to deliver high quality video together with access to the Internet and regular telephone services. Because of the distance limitation, it is envisioned that it will be used to deliver information from a cabinet in the street which is connected to an access network via optical fibers.

Recently, ATM Passive Optical Networks (APON) have emerged as an alternative to providing high-speed access over fiber. APONs were standardized by ITU-T in 1998 (ITU-T Recommendation G.983.1). Also, the *Full Service Access Network* (FSAN) consortium identified the APON as the most cost-effective architecture for high-speed access networks. The FSAN consortium is a group of telecommunication operators and manufacturers which works towards developing a consensus on the systems required to deliver a full set of telecommunication services over an access network. The APON, as its name implies, supports the ATM architecture. It can provide high-speed access to the Internet, voice over packet, and video services equivalent to those provided by a cable operator.

An APON consists of *Optical Network Units* (ONU) which are connected to an *Optical Line Terminator* (OLT) via a passive optical network. The OLT is connected to an ATM network, and it provides and receives traffic to/from the ONUs. The passive optical network is a point-to-multipoint network, with the OLT as the root and the ONUs as the leaves. As shown in Figure 9.2, there are N different fibers fanning out from the OLT. Each fiber is split into multiple fibers using passive optical splitters, shown as circles, so that it can support a maximum of 64 ONUs. The maximum length between the OLT and an ONU is 30 miles.

There are several configurations, depending upon where the ONUs are located. If each ONU is located inside the home, then the configuration is known as *Fiber To The Home* (FTTH). Alternatively, the ONUs may be located in a cabinet in the street, or in the basement of a building. Interconnectivity between the ONU and the homes can be provided using VDSL over copper. This configuration is known as *Fiber To The Curb* (FTTC), *Fiber To The Basement* (FTTB) or *Fiber To The Cabinet* (FTTCab). These configurations differ only in the way in which they are implemented, and from the point of view of the G.983.1 standard, they are all the same.

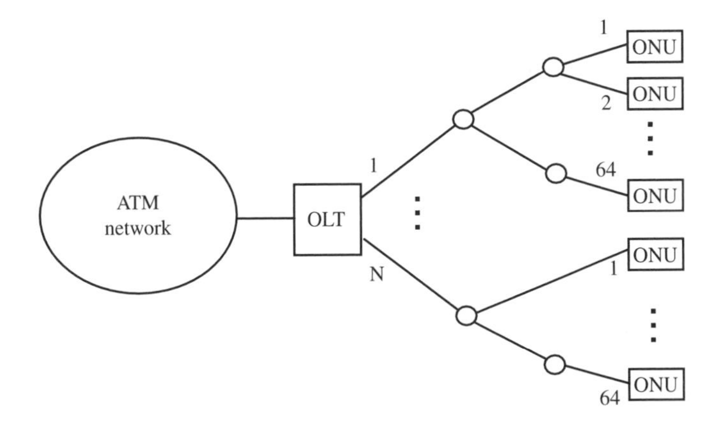

Figure 9.2 The APON architecture (FTTH).

The point-to-multipoint passive optical network supports bidirectional transmission using *Wavelength Division Multiplexing* (WDM). Two wavelengths are used, one for downstream transmission and one for upstream transmission. In the downstream direction, the operating wavelength range on a single fiber is from 1480–1580 nm. In the upstream direction, the operating wavelength range is from 1260–1360 nm. The optical signal transmitted by the OLT is propagated on all *N* fibers, and at each splitter, the incoming optical signal is split as many times as the number of outgoing fiber. Eventually, all the ONUs will receive the same optical signal that was transmitted by the OLT. A time division multiplexing scheme is used that enables the APON user to transmit upstream. Vendors may utilize additional wavelengths for video services.

The transmission rates in the APON could be symmetric or asymmetric. In the symmetric case, the nominal rate is 155 Mbps in both the downstream and upstream direction. In the asymmetric case, the nominal rates are 622 Mbps in the downstream direction and 155 Mbps in the upstream direction.

9.2 THE ADSL TECHNOLOGY

The Asymmetric Digital Subscriber Line (ADSL) technology utilizes the existing twisted pair from the central office to the home to transport data in addition to the basic telephone services. It was originally designed to provide video on demand services transported over switched DS-1 or E1. This type of traffic is referred to in the ADSL standard as the *Synchronous Transfer Mode* (STM) traffic. In its current standard (ANSI T1.413 issue 2 or ITU-T G.992.1), full rate ADSL has been defined to carry either ATM or STM traffic, or both. ADSL is primarily used for ATM traffic, and there is a limited number of applications for STM traffic.

As its name implies, ADSL provides asymmetrical data rates, with the downstream rate being considerably higher than the upstream rate. The data rate depends on the length of the twisted pair, the wire gauge, presence of bridged taps, and cross-couple interference. Ignoring bridged taps, currently ADSL can deliver a full DS-1 or E1 signal downstream over a single unloaded 24 gauge twisted pair for a maximum distance of 18 000 feet. Up to 6.1 Mbps is possible for a maximum distance of 12 000 feet, and 8.128 Mbps for a maximum distance of 9000 feet. Upstream data rates are presently in the 64–800 Kbps range.

The deployment of ADSL over the twisted pair requires an ADSL transmission unit at either end of the line. The ADSL transmission unit at the customer premises is referred to as the *ADSL Transceiver Unit, Remote terminal* (ATU-R), and the ADSL transmission unit at the central office is referred to as the *ADSL Transceiver Unit, Central Office* (ATU-C).

The signal that arrives at the ATU-R over the twisted pair contains both ADSL data and voice. It is necessary to split this signal and deliver the voice signal to the telephone sets and the ADSL signal to a PC. This can be achieved using a splitter, as shown in Figure 9.3(a). This solution utilizes the fact that category 3 or 5 telephone wires used in a home typically consist of two sets of wires (they may also consist of three sets of wires). Voice is usually carried on the inner pair. The splitter splits the incoming signal so that the voice is directed to the inner pair and the ADSL data to the outside pair. This is achieved using passive elements that simply perform a low pass filter that directs voice signals to the inner pair. The high frequency ADSL traffic is coupled via a transformer to the outer pair. A phone, fax machine or answering machine can be plugged into any telephone

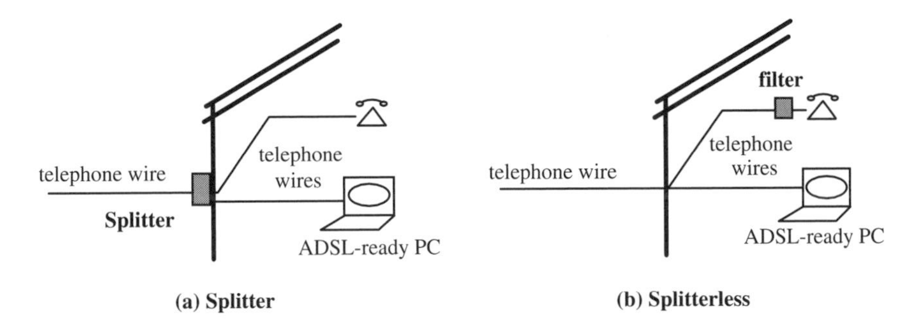

Figure 9.3 Two possible solutions at the customer's premises.

plug in the house. The PC used to access the Internet has to have a built-in or external ATU-R that permits it to receive and transmit ADSL data, and it can be plugged into any telephone plug. If the quality of the twisted pair in a home is not good, a dedicated high-quality line is installed from the splitter to the ATU-R.

The installation of the splitter requires a visit by a technician. An alternative solution, is to use *splitterless* ADSL, as shown in Figure 9.3(b). Splitterless ADSL does not require installation by a technician, and consequently it is cheaper to deploy than the ADSL modem that does require a splitter. In this case, the signal transmitted over the twisted pair, which contains both ADSL data and voice, is propagated throughout the home telephone wires. The voice signal is filtered out using a high pass filter inside the ATU-R. On the other hand, the ADSL signal can cause a strong audible noise through the telephone set. Therefore, each phone is attached to a telephone plug through a filter, which filters out the ADSL signal and, at the same time, isolates voice events such as ring and on/off hook from the ADSL signal. The PC can be plugged in to any telephone plug.

The splitterless ADSL standard is referred to as G.LITE (ITU-T recommendation G.992.2). G.LITE is low-speed, splitterless ADSL solution, with a downstream data rate of 1.536 Mbps and an upstream rate of 512 Kbps. Current developments in silicon migration, better power management techniques, and splitterless implementation has permitted the full rate ADSL to be also used in a splitterless mode. As a result, in most countries the G.LITE solution never became popular.

Now, let us take a look at the ATU-C at the central office. As shown in Figure 9.4, in the downstream direction the voice signal is added after the ADSL signal leaves the ATU-C. In the upstream direction, the voice signal is extracted from the ADSL signal, before the ATU-C. The ATU-C generates the ADSL signal in the downstream direction, and terminates the ADSL signal in the upstream direction. A number of ATU-Cs are serviced by an *ADSL access multiplexer*, known as DSLAM, which provides connectivity to IP and ATM networks.

The DSLAM is in fact an ATM switch. It typically has an OC-3 or higher link to an ATM access backbone network, and it also has ADSL links serving a number of customer premises. Each ADSL link at the DSLAM is associated with an ATU-C, which is the physical layer associated with the link (see Section 4.5).

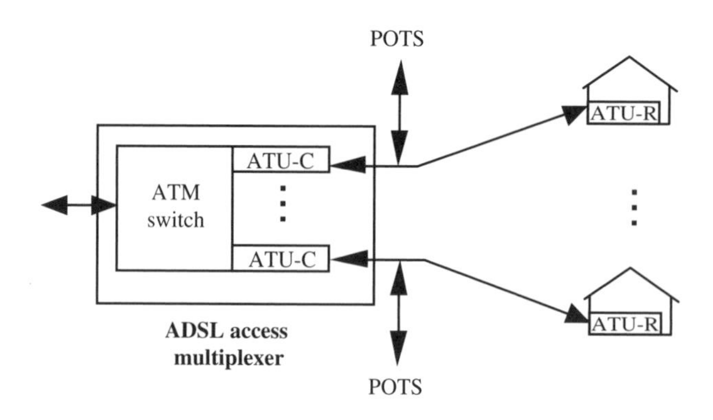

Figure 9.4 The ADSL access multiplexer (DSLAM).

We now describe how an ATU-C or an ATU-R works. The protocols used to provide IP and ATM services over ADSL are described in Section 9.3.

9.2.1 The Discrete Multi-Tone (DMT) Technique

The *Discrete Multi-Tone* (DMT) technology is the standardized line coding technique used for ADSL. DMT devices can easily adjust to changing line conditions, such as moisture or interference, and it is resistant to noise and the presence of digital signals or adjacent wire pairs (cross-talk).

In the DMT technique, the entire bandwidth of the twisted pair is divided into a large number of equally spaced subchannels, also known as *tones*. The twisted pair's bandwidth extends to 1.1 MHz, and it is divided to 256 subchannels, each occupying 4.3125 KHz. The lower subchannels 1–6 are reserved for the voiceband region, and are used to provide basic telephone services. The remaining subchannels are used by ADSL.

ADSL is bidirectional, which means that both the upstream and downstream data is sent over the same twisted pair. In ADSL, bidirectional transmission over the twisted pair can be implemented using either *Frequency Division Multiplexing* (FDM) or echo cancellation. In FDM, there are up to 32 upstream subchannels (i.e. from the customer premises to the network) occupying the frequencies immediately above the voiceband region. Also, there are up to 218 downstream subchannels (i.e. from the network to the customer premises) occupying the frequencies above the upstream subchannels. An alternative solution is to let the upstream and downstream subchannels use the same frequencies, and separate them using echo cancellation. Not all subchannels are used for the transfer of information; some are used for network management and performance monitoring. All subchannels are monitored constantly for performance and errors, and the speed of each subchannel or group of channels can actually vary with a granularity of 32 Kbps.

Transmission is achieved by dividing time into fixed-sized intervals. Within each interval, DMT transmits a data frame which consists of a fixed number of bits. The bits in a data frame are divided into groups of bits, and each group is transmitted over a different subchannel. The number of bits sent over each subchannel can be varied depending upon the signal and noise level in each subchannel. Using the *Quadrature Amplitude Modulation* (QAM) technique, the bits allocated to each subchannel are converted into a complex number which is used to set the subchannel's amplitude and phase for the interval. The signals are all added up and sent to the twisted pair. This signal resulting from each data frame is known as the *DMT symbol*.

9.2.2 Bearer Channels

A diagram of the ATU-R showing the flow of data in the downstream direction is given in Figure 9.5. The flow of data in the upstream direction in the ATU-C has a similar structure. The data transported by an ATU-R or ATU-C is organized into seven independent logical bearer channels. Of these seven channels, four are unidirectional channels from the network to the customer premises. These four channels are referred to as the *simplex bearer channels*, and they are designated as AS0, AS1, AS2 and AS3. The remaining three channels are duplex, and they are referred to as the *duplex bearer channels*. They are bidirectional channels between the network and the customer premises, and they are designated as LS0, LS1 and LS2. The three duplex bearer channels may also be configured as independent unidirectional simplex channels. All bearer channels can be programmed to transmit at a speed which is an integer multiple of 32 Kbps, as shown in Table 9.2. The maximum total data rate of the ADSL system depends upon the characteristics of the twisted pair on which the system is deployed.

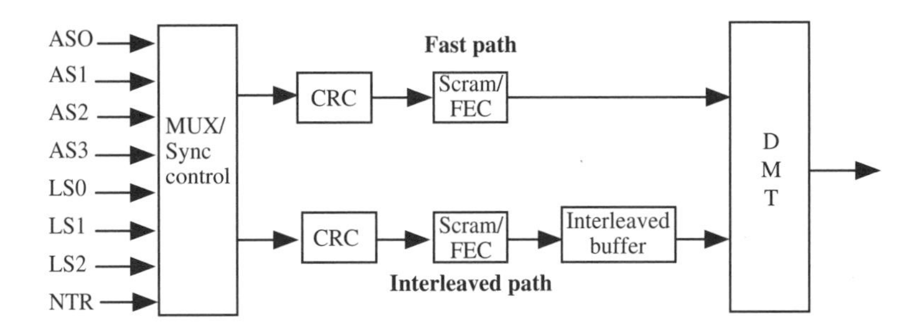

Figure 9.5 The fast path and the interleaved path in ATU-R.

Table 9.2 Data rates for the bearer channels.

Bearer channel	Lowest required multiple	Largest required multiple	Corresponding highest data rate (Kbps)
AS0	1	192	6144
AS1	1	144	4608
AS2	1	96	3072
AS3	1	48	1536
LS0	1	20	640
LS1	1	20	640
LS2	1	20	640

STM traffic is mapped in bearer channels AS0 and LS0 in the downstream direction, and in LS0 in the upstream direction. Other bearer channels can also be provisioned. ATM traffic is mapped in the downstream direction in bearer channel AS0 and in LS0 in the upstream direction. Other bearer channels can also be provisioned.

Some applications running at the customer premises may require a reference clock. In view of this, in addition to transporting user data, an ATU-C may optionally transport a *Network Timing Reference* (NTR) to an ATU-R.

A bearer channel in an ATU-R or ATU-C can be assigned either to the *fast path* or the *interleaved path*. The two paths in the ATU-R are shown in Figure 9.5. The fast path provides low delay, whereas the interleaved path provides greater delay but lower error rate. CRC, Forward Error Correction (FEC) and scrambling can be applied to each path. Bearer channel AS0 carrying downstream ATM traffic can be transmitted over either the fast path or the interleaved path. Upstream ATM data are transmitted in LS0 either over the fast path or the interleaved path. The bearer channels carrying STM traffic in either direction are transmitted over either the fast path or the interleaved path. The choice between the fast path and the interleaved path in the downstream direction may be independent of that in the upstream direction

9.2.3 The ADSL Super Frame

As mentioned above, a data frame consists of a fixed number of bits, and it is transmitted every fixed interval using the DMT technique. Each data frame combines bits interleaved from the fast and the interleaved paths. The data frames are combined into a super frame consisting of 68 data frames plus a synchronization data frame, as shown in Figure 9.6. Each data frame is transmitted on the twisted pair as a DMT symbol. The rate of transmission of DMT symbols is 4000 symbol/s. Since a synchronization data frame is transmitted for every 68 data frames, the transmission rate on the twisted pair is actually slightly higher, i.e. (69/68)*4000 symbols/s. That is, the super frame repeats every 17 ms.

9.3 SCHEMES FOR ACCESSING NETWORK SERVICE PROVIDERS

Network Service Providers (NSP) are content providers, Internet service providers and corporate networks. Providing access to NSPs is an important service to the ADSL users.

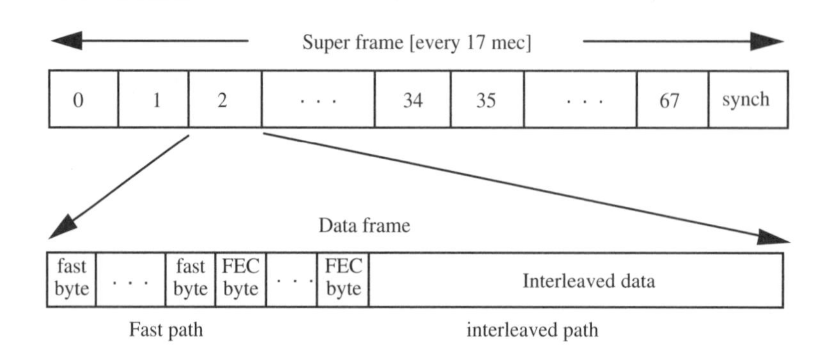

Figure 9.6 The super frame.

In this section, we describe two different schemes that can be used to provide connectivity to network service providers, namely the *L2TP access aggregation* scheme and the *PPP terminated aggregation* scheme.

The ADSL Forum's reference architecture, shown in Figure 9.7, defines the connectivity between ADSL users and NSPs. This reference architecture consists of customer premises, an access network, a regional public network and NSPs. The customer premises may be a residence, a home office or a small business office. At a customer premises, there may be one or more computers interconnected by a network. The access network includes the ATU-Rs at the customer premises, the DSLAMs that serve these ATU-Rs, and an access backbone network that interconnects all the DSLAMs. Connectivity to NSPs and to a *Regional Operations Center* (ROC) is provided through a regional public network. Typically, the access network is managed by a telephone operator, who controls it via an ROC. The telephone operator may be the local telephone operator, known as the *Incumbent Local Exchange Carrier* (ILEC), or a national or newcomer telephone operator, known as the *Competitive Local Exchange Carrier* (CLEC).

In most cases, the ATU-Rs do not have the necessary functionality to set up SVCs. Instead, PVCs are used. Providing each ATU-R with a PVC to each NSP requires a large number of PVCs to be set up and managed. A more scaleable approach is to provide a *Network Access Server* (NAS), as shown in Figure 9.7. The role of the NAS is to terminate all the PVCs from the ATU-Rs, and then aggregate the traffic into a single connection for each NSP.

ADSL users set up sessions to NSPs using the *Point-to-Point Protocol* (PPP). This protocol was designed to provide a standard method for transporting datagrams from different protocols over a full-duplex link. PPP provides a number of functions, such as assignment of an IP address by a destination network, domain name auto-configuration,

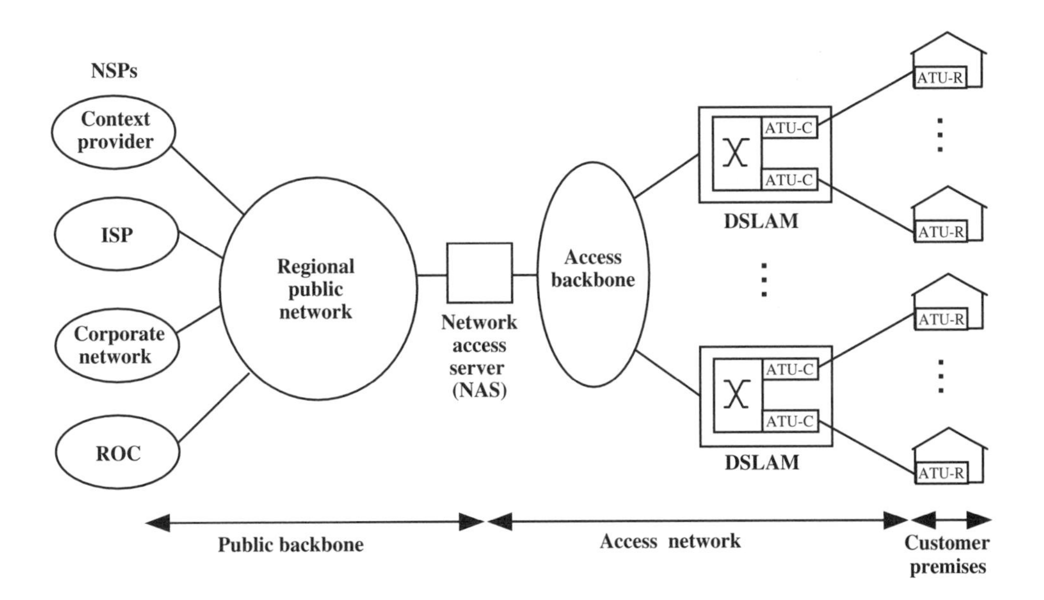

Figure 9.7 The ADSL reference architecture.

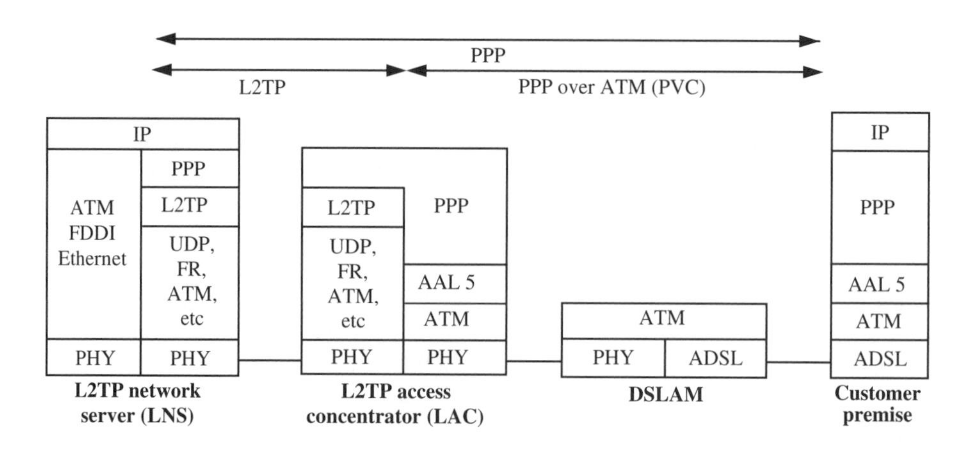

Figure 9.8 The L2TP access aggregation scheme.

multiplexing of different network layer protocols, authentication, encryption, compression and billing. PPP frames are transported using a default HDLC-like encapsulation. When PPP runs on top of ATM, PPP frames are mapped into AAL 5 PDUs using either the *VC-multiplexed PPP* scheme or the *LLC encapsulated PPP* scheme. In the former scheme, a PPP frame is directly carried in an AAL 5 PDU. In the latter scheme, a PPP frame is also carried in an AAL 5 PDU after it is further encapsulated with a 2-byte LLC header and a 1-byte network layer protocol identifier.

9.3.1 The L2TP Access Aggregation Scheme

This scheme is based on IETF's *Layer 2 Tunneling Protocol* (L2TP). The protocol stacks involved in this scheme are shown in Figure 9.8. For simplicity, we assume that an ADSL user at a customer premises is a single computer, rather than a network of computers interconnected via an ATM network. An ADSL user is connected to the DSLAM over ADSL, and the DSLAM is connected to an NAS, referred to as the *L2TP Access Concentrator* (LAC), over an ATM network. Finally, the LAC is connected to the *L2TP Network Server* (LNS) of each NSP over a network such as IP, frame relay and ATM.

The ADSL user is connected to the LAC with an ATM PVC via the DSLAM. This connection uses AAL 5. The LAC and the LNS of an NSP are connected by an L2TP *tunnel*. An L2TP tunnel is not an actual connection in the sense of an ATM connection, rather it is a logical connection between the L2TP on the LAC and its peer L2TP on the LNS. A PPP session between the ADSL user and the LNS is established as follows. The ADSL user sends a request to the LAC over AAL 5 to initiate a PPP session to an LNS. This request is forwarded by the LAC to the LNS over an L2TP tunnel. Once the PPP session is established, IP packets can begin to flow between the ADSL user and the LNS.

A tunnel between the LAC and an LNS can multiplex several PPP sessions, each associated with a different ADSL user. Also, there may be several tunnels between the LAC and an LNS. L2TP utilizes two types of messages, namely, *control messages* and *data messages*. Control messages are used to establish, maintain and clear tunnels and PPP sessions on demand. Data messages are used to carry PPP frames over a tunnel.

The control and data messages are encapsulated with a common L2TP header. Some of the fields in this header are: type bit (T), length bit (L), priority bit (P), sequence bit (S), length, tunnel ID, session ID, sequence number (Ns), and expected sequence number (Nr). The type bit field indicates whether the L2TP packet is a control or a data message. The length bit field indicates whether the length field is present. If it is present, the length field gives the total length of the L2TP packet in bytes. The priority bit is used for data messages. If it is set to 1, then the L2TP packet is to be given preferential treatment within the L2TP queues. The L2TP packet is associated with a tunnel ID and a PPP session ID, given in the tunnel IP field and the session ID field, respectively. The sequence bit indicates whether sequence numbers are being used. If they are used, then they are carried in the Ns and Nr fields, which are similar to the N(R) and N(S) fields in the HDLC header (see Section 2.4). That is, the Ns field contains the sequence number of the transmitted L2TP packet, and the Nr field contains the next sequence number the transmitting L2TP expects to receive from its peer L2TP.

A reliable channel between two L2TP peers is provided by L2TP for control messages only. The Ns and Nr sequence numbers are used to detect out-of-sequence packets and missing packets. Lost packets are recovered by retransmission. Data messages may use sequence numbers optionally to reorder packets and detect lost packets. However, no retransmission of data messages takes place. L2TP runs over a network such as IP using UDP, frame relay and ATM.

The establishment of a session within a tunnel is triggered when the LAC receives a request from an ADSL user to initiate a PPP session to an NLS. Each session within a tunnel corresponds to a single PPP session. Once the session is established, PPP frames can flow between the ADSL user and the LNS. Specifically, PPP frames are transmitted to the LAC over the ATM PVC. The LAC receives the PPP frames from AAL 5, encapsulates each frame with an L2TP header, and transmits it to the LNS as a data message.

9.3.2 The PPP Terminated Aggregation Scheme

This scheme is based on the *Remote Authentication Dial In User Service* (RADIUS) protocol. This is a client/server protocol used for authentication, authorization and accounting. An NAS, referred to as the *Broadband Access Server* (BAS), acts as a client to a RADIUS server, which is managed by an NSP. The BAS is responsible for passing authentication information such as user login and password to the RADIUS server. This authentication information is submitted by an ADSL user when it initiates a PPP session. The RADIUS server is responsible for authenticating the ADSL user, and then returning configuration information necessary for the BAS to deliver service to the ADSL user. A BAS also sends the RADIUS server accounting information.

The protocol stacks involved in this scheme are shown in Figure 9.9. As in the previous scheme, we assume that an ADSL user at a customer premises is a single computer, rather than a network of computers interconnected via an ATM network. The ADSL user is connected to the DSLAM using ADSL. On the side of the access backbone network, the DSLAM is connected to the BAS via an ATM network. Finally, the BAS is connected to NSP routers over a network, such as IP, frame relay and ATM.

An ADSL user is connected to the BAS with an ATM PVC via the DSLAM. A PPP session initiated by an ADSL user terminates at the BAS, instead of being tunneled to the NSP, as in the previous scheme. The BAS sends the user authentication information to the

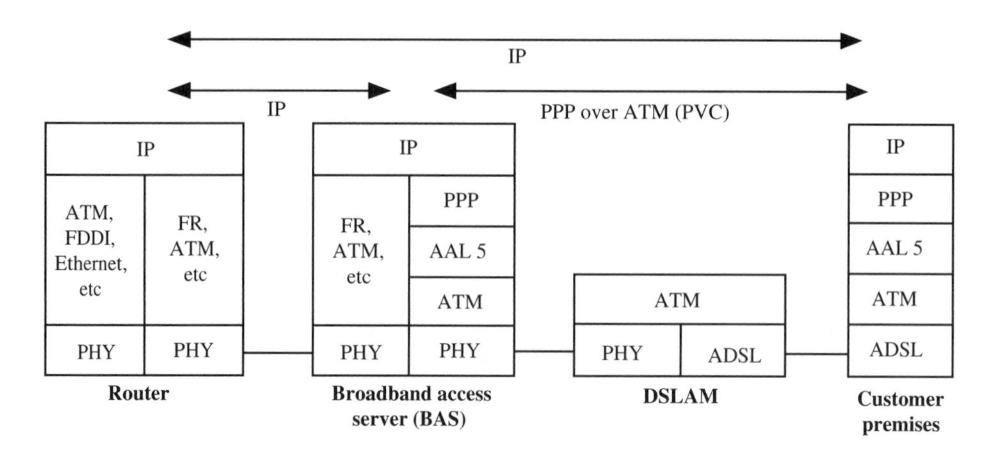

Figure 9.9 The PPP terminated aggregation scheme.

appropriate RADIUS server, and the PPP session is established after the RADIUS server validates the user. The ADSL user can now transmit IP packets, which are forwarded by the BAS to the router of the appropriate NSP.

PROBLEMS

1. You need to download a 20 MB file to your computer at home.
   (a) How long does it take to download it assuming that you are connected with a 56 K modem, that gives a throughput of 49 Kbps?
   (b) How long does it take to download the same file assuming that your computer is connected to an ADSL modem, which provides a throughput of 1 Mbps?
2. Explain why splitterless ADSL modem is preferable over the ADSL modem that requires a splitter.
3. What is the difference between the fast path and the interleaved path in an ADSL modem?
4. What is the advantage of using the Network Access Server (NAS)?
5. In L2TP, why are control messages transmitted over a reliable channel and not data messages?

Part 4

Signaling in ATM Networks

10

Signaling over the UNI

In this chapter, we describe the signaling protocols that are used to establish a point-to-point SVC and a point-to-multipoint SVC over the private UNI. ITU-T recommendation Q.2931 is used to establish a point-to-point VC connection, and ITU-T recommendation Q.2971 is used to establish a point-to-multipoint VC connection. Both signaling protocols run on top of a specialized AAL, known as the *Signaling AAL* (SAAL). A special sublayer of this AAL is the *Service-Specific Connection Oriented Protocol* (SSCOP).

We first describe the main features of SAAL and SSCOP, and present the various ATM addressing schemes. Then, we discuss in detail the signaling messages and procedures used by Q.2931, Q.2971, and the *Leaf Initiated Join* (LIJ) capability.

10.1 CONNECTION TYPES

We recall that in ATM networks there are two types of connections, Permanent Virtual Connections (PVC) and Switched Virtual Connections (SVC). PVCs are established using network management procedures, whereas SVCs are established in real-time using signaling procedures. In general, PVCs remain established for long periods of time, whereas SVCs remain active for an arbitrary amount of time. PVCs and SVCs may be point-to-point, point-to-multipoint and multipoint-to-multipoint.

A point-to-point virtual circuit connection is bidirectional, and is composed of two unidirectional connections, one in each direction. Both connections are established over the same physical route. Bandwidth requirements and quality of service may be specified separately for each direction. This type of connection is defined by ITU-T as *type 1*. Point-to-point connections can be established over the private UNI using the signaling protocol Q.2931.

A *type 2* virtual circuit connection is a unidirectional point-to-multipoint connection. It consists of an ATM end device, known as the *root*, which transmits information to a number of other ATM end devices, known as the *leaves*. Signaling is provided to establish a type-2 connection, and also to add/remove leaves on demand. A type 2 connection is established using Q.2971 in conjunction with Q.2931. Specifically, the connection to the first leaf is established using Q.2931. Q.2971 signaling procedures are then used to add new leaves or drop existing leaves. Q.2971 was designed so that a leaf can only be added or dropped by the root of the connection. In addition to Q.2971, the ATM Forum has introduced the *Leaf Initiated Join* (LIJ) capability, which permits a leaf to add or drop a point-to-multipoint VC connection without intervention from the root.

We note that both Q.2931 and Q.2971 have been modified by the ATM Forum. For simplicity, we refer to these modified protocols by their original ITU-T recommendation names, Q.2931 and Q.2971.

A *type 3* connection is a unidirectional, multipoint-to-point connection, used primarily in a residential environment. Finally, a *type 4* connection is a multipoint-to-multipoint connection that can be used for conferencing. An equivalent connection to type 4 can be set up by allowing each ATM end device to set up its own point-to-multipoint connection.

A call, whether point-to-point or point-to-multipoint, may require multiple connections, each with a different quality of service. This type of connection is useful for multimedia calls which combine videoconferencing, file transfers and a shared white board. Each part of the application may be carried by a different connection. This could be accomplished by setting up one connection at a time using the appropriate signaling protocol. Alternatively, these connections can be established as a group in a single request using appropriate signaling. Such signaling is currently under development.

10.2 THE SIGNALING PROTOCOL STACK

The signaling protocol stack is shown in Figure 10.1. It is analogous to the ATM protocol stack shown in Figure 4.5 (see Section 4.3). The ATM protocol stack shows the protocol layers used for the transfer of data, whereas the stack in Figure 10.1 shows the protocol layers used for setting-up an SVC. As we can see, a signaling protocol such as Q.2931, Q.2971 and PNNI is an application that runs on top of a specialized ATM adaptation layer known as the *Signaling AAL* (SAAL). Below SAAL, we have the familiar ATM layer and the physical layer. The signaling protocol stack is often referred to as the *control plane*, as opposed to the *data plane*, that refers to the ATM protocol stack.

10.3 THE SIGNALING ATM ADAPTATION LAYER (SAAL)

As shown in Figure 10.2, SAAL consists of an SSCS, which is composed of the *Service-Specific Coordination Function* (SSCF) and the *Service-Specific Connection Oriented Protocol* (SSCOP). The common part of the convergence sublayer is AAL 5, described in Chapter 5. There is no SAR layer. The SSCF maps the services required by the signaling protocol to the services provided by SSCOP. The SSCOP is a protocol designed to provide a reliable connection over the UNI to its peer SSCOP. This connection is used by a signaling protocol to exchange signaling messages over the UNI with its peer protocol.

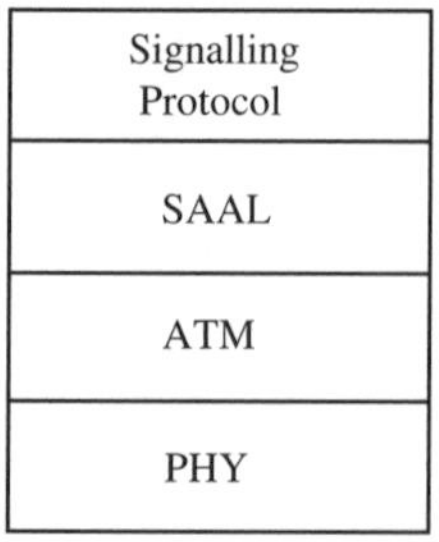

Signalling Protocol
SAAL
ATM
PHY

Figure 10.1 The signaling protocol stack.

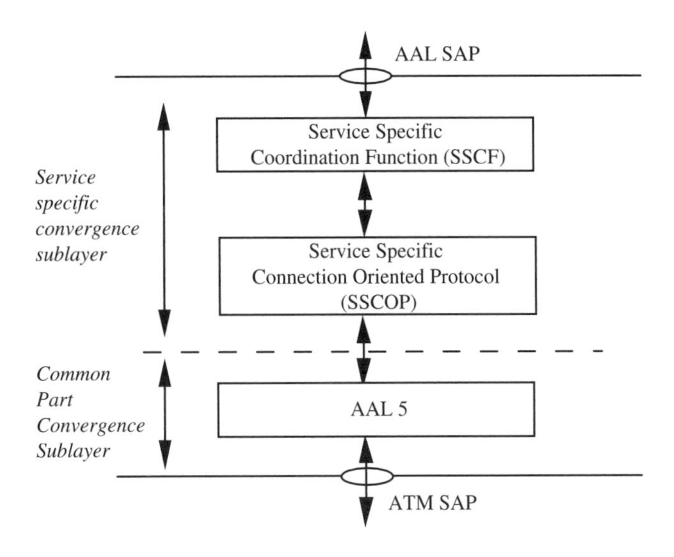

Figure 10.2 The signaling AAL (SAAL).

10.3.1 The SSCOP

SSCOP provides most of the services provided by LAP-D in ISDN and *Message Transfer Part* (MTP) level 2 in *Signaling System no. 7* (SS7). It was designed for ATM networks with a high bandwidth-delay product, i.e. the transmission speed of an ATM link and the propagation time are both high. In such an environment, the traditional ARQ schemes described in Section 2.3 are not very effective.

SSCOP transfers data to its peer SSCOP in a variable-length PDU, referred to as a *frame*. SSCOP's main function is to establish and release a connection to a peer SSCOP, and to maintain an assured transfer of frames over the connection. This is done using the frames given in Table 10.1. The establishment and release of a connection is achieved using the BEGIN and the END frames described in Table 10.1.

The assured transfer of frames is achieved using error detection and recovery by retransmission, and flow control which is based on an adjustable sliding window. The data is carried in SEQUENCED DATA (SD) frames. An SD frame uses a sequence number, and it can carry a variable length payload of up to 65 535 bytes. The frames STATUS REQUEST (POLL), SOLICITED STATUS RESPONSE (STAT) and UNSOLICITED STATUS RESPONSE (USTAT) are used to implement a retransmission scheme for erroneously received SD frames or lost SD frames. Specifically, the transmitter periodically sends a POLL frame to request the status of the receiver. This POLL is either triggered when a timer expires, or after a certain number of SD frames has been sent. The POLL contains the sequence number of the next SD frame to be sent by the transmitter and a poll sequence number, which essentially functions as a time-stamp. The receiver, upon receipt of a POLL, responds with a STAT frame which contains the following information: an SD sequence number up to which the transmitter may transmit (i.e. the window), the number of the next SD frame expected, the echoed poll sequence number, and a list of all SD frames that are currently missing or have been received erroneously. The receiver can determine the missing SD frames by checking for gaps in its buffer, and by examining

Table 10.1 The SSCOP frames.

Name	Description
BEGIN	Used to initially establish an SSCOP connection or to re-establish an existing SSCOP connection.
BEGIN ACKNOWLEDGE	Used to acknowledge acceptance of an SSCOP connection request by the peer SSCOP.
END	Used to release an SSCOP connection between two peer SSCOP entities.
END ACKNOWLEDGE	Used to confirm the release of an SSCOP connection requested by the peer SSCOP.
BEGIN REJECT	Used to reject the establishment of a connection requested by the peer SSCOP.
RESYNCHRONIZE	Used to resynchronize the buffer and the data transfer state variables in the transmit direction of a connection.
RESYNCHRONIZE ACKNOWLEDGE	Used to acknowledge the resynchronization of the local receiver in response to the resynchronize frame.
SEQUENCED DATA (SD)	Used to transfer sequentially numbered frames containing user information.
STATUS REQUEST (POLL)	Used by transmitting SSCOP to request status information from the receiving SSCOP.
SOLICITED STATUS RESPONSE (STAT)	Used to respond to a POLL. It contains the sequence numbers of outstanding SD frames and credit information for the sliding window.
UNSOLICITED STATUS RESPONSE (USTAT)	Similar to STAT message, but issued by the transmitter when a missing or erroneous frame is identified.

the SD sequence number contained in the POLL. Based on the received STAT message, the transmitter retransmits the outstanding SD frames and advances the transmit window.

If the receiver detects an erroneous or missing SD frame, it sends a USTAT, instead of having to wait for a POLL. A USTAT frame is identical to a STAT frame, except that it is not associated with a POLL. The USTAT frame can also be used by the receiver to ask the transmitter to increase or decrease the frequency of POLL frames.

10.3.2 Primitives

SAAL functions are accessed by a signaling protocol, such as Q.2931, Q.2971 and PNNI, through the AAL-SAP, using the following primitives: AAL-ESTABLISH, AAL-RELEASE, AAL-DATA and AAL-UNIT-DATA.

The AAL-ESTABLISH is issued by a signaling protocol to SAAL to request the establishment of a connection over the UNI to its peer protocol. This is necessary for the two peer signaling protocol to exchange signaling messages. This is a reliable connection, and it is managed by the SSCOP as described above.

The AAL-RELEASE primitive is a request by a signaling protocol to SAAL to terminate a connection established earlier on using the AAL-ESTABLISH primitive.

The AAL-DATA primitive is used by a signaling protocol to request the transfer of a signaling message to its peer signaling protocol. Signaling messages have a specific structure, and they will be discussed below in detail. Finally, the AAL-UNIT-DATA is used to request a data transfer over an unreliable connection.

These primitives can be one of the following four types: *request*, *indication*, *response* and *confirm*. These types are shown in Figure 10.3. A request type is used when the signaling protocol wants to request a service from SAAL. An indication type is used by SAAL to notify the signaling protocol of a service-related activity. A response type is used by the signaling protocol to acknowledge receipt of a primitive typed indication. A confirm type is used by SAAL to confirm that a requested activity has been completed.

An example of how these primitives and their types are used to establish a new connection over the UNI between two peer signaling protocols is shown in Figure 10.4. The primitive AAL-ESTABLISH.request is used to request SAAL to establish a connection. (To simplify the presentation, we do not present the signals exchanged between the SSCF and the SSCOP.) In response to this request, SSCOP sends a BEGIN frame to its peer SSCOP. The peer SAAL generates an AAL-ESTABLISH.indication to the peer signaling protocol, and its SSCOP returns a BEGIN ACKNOWLEDGE frame, upon receipt of which the SAAL issues a AAL-ESTABLISH.confirm to the signaling protocol.

An example of how a connection over the UNI between two peer signaling protocols is terminated is shown in Figure 10.5. We can see that the signaling protocol issues

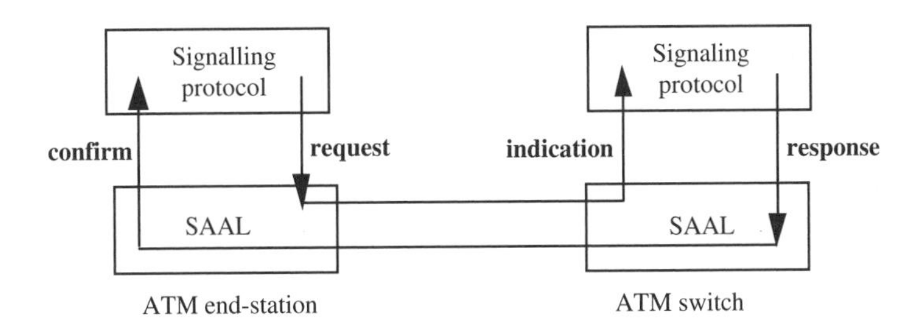

Figure 10.3 The four primitive types.

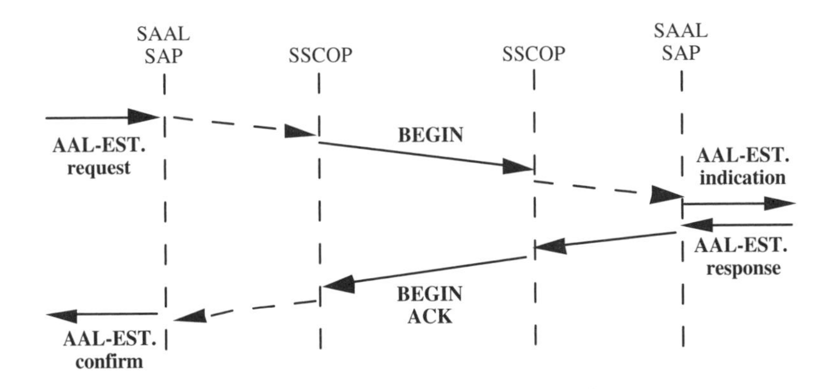

Figure 10.4 Establishment of a connection between two peer signaling protocols.

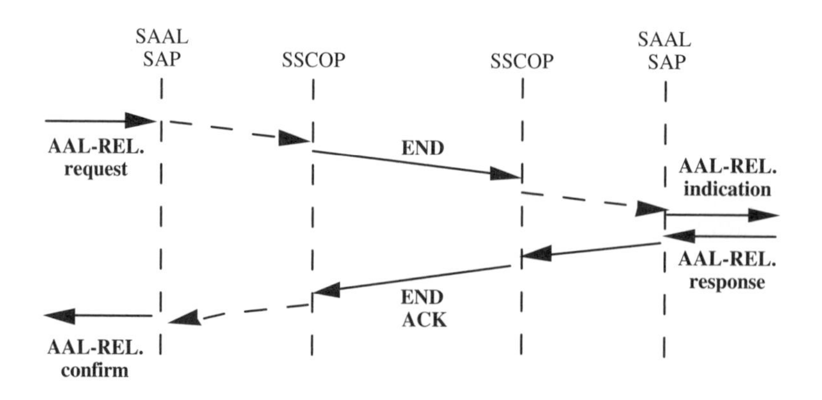

Figure 10.5 Termination of a connection between two peer signaling protocols.

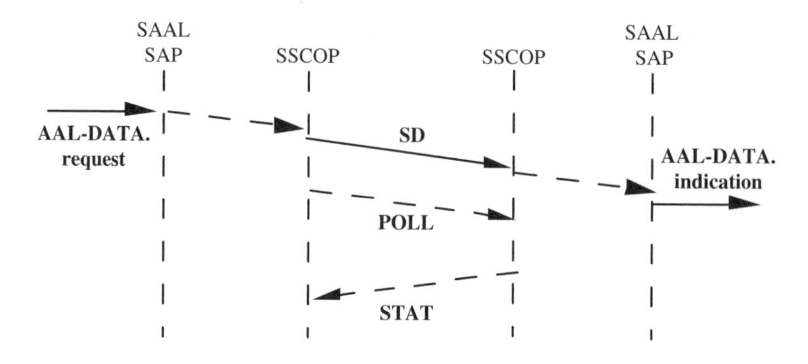

Figure 10.6 Transfer of a signaling message.

an AAL-RELEASE.request to SAAL, in response of which the SSCOP sends an END frame to its peer SSCOP. The peer SAAL sends an AAL-RELEASE.indication to the peer signaling protocol, and its SSCOP returns an END ACKNOWLEDGE frame, upon receipt of which the SAAL issues a AAL-RELEASE.confirm to the signaling protocol.

An example of how a signaling protocol transfers messages to its peer protocol is shown in Figure 10.6. The signaling protocol transfers a message to SAAL in an AAL-DATA.request, which is then transferred by SSCOP in an SD frame. The SD frame is passed onto AAL 5, which encapsulates it and then breaks it up to 48 byte segments, each of which is transferred by an ATM cell. Figure 10.6 also shows the POLL/STAT frames exchanged between the two peer SSCOPs. The SD frame at the destination side is delivered to the peer signaling protocol using the AAL-DATA. indication primitive.

10.4 THE SIGNALING CHANNEL

This is a VC connection that is used exclusively to carry the ATM traffic that results from the exchange of signaling messages between two peer signaling protocols. It is a default connection identified by VPI = 0 and VCI = 5. This signaling channel is used to control VC connections within all the virtual paths. It is also possible to set-up a signaling channel with a VCI = 0 within a virtual path connection with a VPI other than 0, say with a VPI = x. In this case, this signaling channel can only be used to control VC connections within the virtual path x.

The signaling channel VPI/VCI = 0/5 is used in conjunction with the signaling mode known as *nonassociated signaling*. In this mode, all the VC connections are created, controlled and released via the signaling channel VPI/VCI = 0/5. A signaling channel within a VPI = x, where x > 0, is used in conjunction with the signaling mode known as *associated signaling*. In this mode, only the VC connections within the virtual path x are created, controlled and released via the signaling channel VPI/VCI = x/5.

10.5 ATM ADDRESSING

Each ATM end device and each ATM switch has a unique ATM address. Private and public networks use different ATM addressing formats. Public ATM networks use E.164 addresses, whereas ATM private network addresses use the OSI *Network Service Access Point* (NSAP) format.

The E.164 addressing scheme is based on the global ISDN numbering plan, shown in Figure 10.7. It consists of 16 digits, each coded in Binary Coded Decimal (BCD) using 4 bits. Thus, the total length of the E.164 address is 64 bits, or 8 bytes. The first digit indicates whether the address is a unicast or multicast. The next three digits indicate the country code, and the remaining digits are used to indicate an area or city code, an exchange code, and an end device identifier. When connecting a private ATM network to a public network, only the UNIs connected directly to the public network have an E.164 address.

The private ATM addresses are based on the concept of hierarchical addressing domains, and they are 20 bytes long. As shown in Figure 10.8, the address format consists of two parts, namely, the *Initial Domain Part* (IDP) and the *Domain-Specific Part* (DSP).

The IDP specifies an administration authority which has the responsibility for allocating and assigning values for the DSP. It is subdivided to the *Authority and Format Identifier* (AFI) and the *Initial Domain Identifier* (IDI). AFI specifies the format of the IDI, and the abstract syntax of the DSP field. The length of the AFI field is one byte. The IDI specifies the network addressing domain, from which the DSPs are allocated and the network addressing authority responsible for allocating values of the DSP from that domain. The following three IDIs have been defined by the ATM Forum:

1. DCC (Data Country Code): this field specifies the country in which the address is registered. The country codes are specified in ISO 3166. These addresses are administered by the ISO's national member body in each country. The length of this field is two bytes, and the digits of the data country code are encoded using BCD.

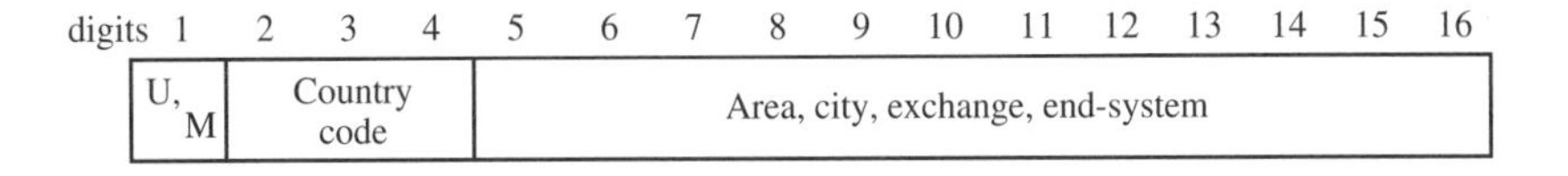

Figure 10.7 The E.164 addressing scheme.

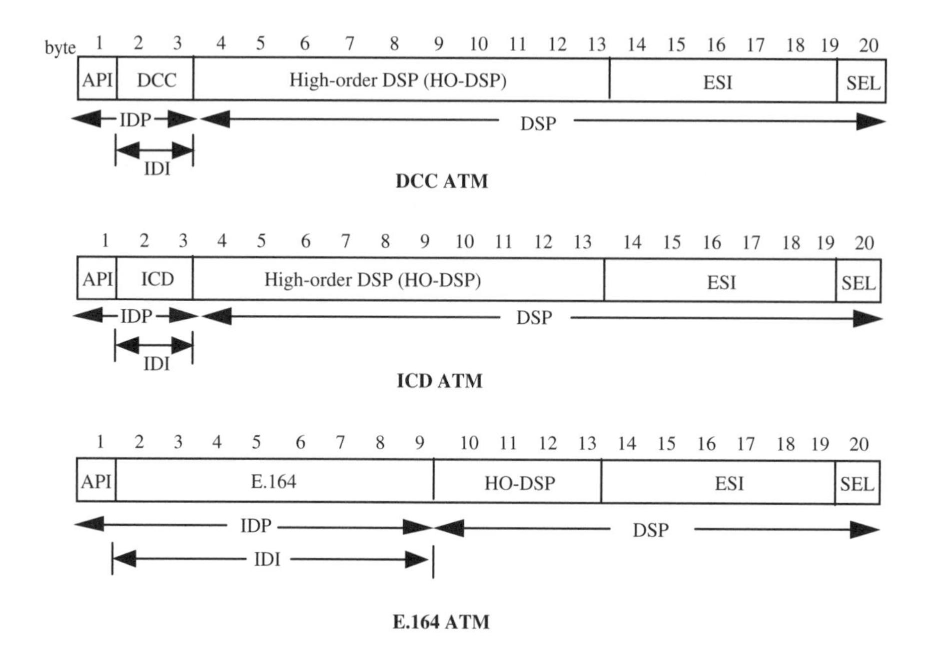

Figure 10.8 The NSAP ATM formats.

2. ICD (International Code Designator): the ICD field identifies an authority which administers a coding scheme. This authority is responsible for the allocation of identifiers within this coding scheme to organizations. The registration authority for the international code designator is maintained by the British Standards Institute. The length of the field is two bytes and the digits of the international code designator are encoded using BCD.
3. E.164 addresses.

The DSP field consists of the *High-Order DSP* (HO-DSP) field, the *End System Identifier* (ESI) field and the SEL (selector) field. The coding for the HO-DSP field is specified by the authority or the coding scheme identified by the IDP. The authority determines how identifiers will be assigned and interpreted within that domain. The authority can create further subdomains. That is, it can define a number of sub-fields of the HO-DSP and use them to identify a lower authority, which in turn defines the balance of HO-DSP. The content of these subfields describes a hierarchy of addressing authorities, and conveys a topological structure.

The End System Identifier (ESI) is used to identify an end device. This identifier must be unique for a particular value of the IDP and HO-DSP fields. The ESI can also be globally unique by populating it with a 48-bit IEEE MAC address. Finally, the selector (SEL) field has only local significance to the end device, and it is not used in routing. It is used to distinguish different destinations reachable at the end device.

Of interest is how IP addresses can be mapped to the NSAP structure. Figure 10.9 shows a mapping which can eliminate the use of ATMARP.

To show how the NSAP ATM addresses can be used, we describe the ATM addressing scheme of the private ATM network NCANet (the North Carolina Advanced Network).

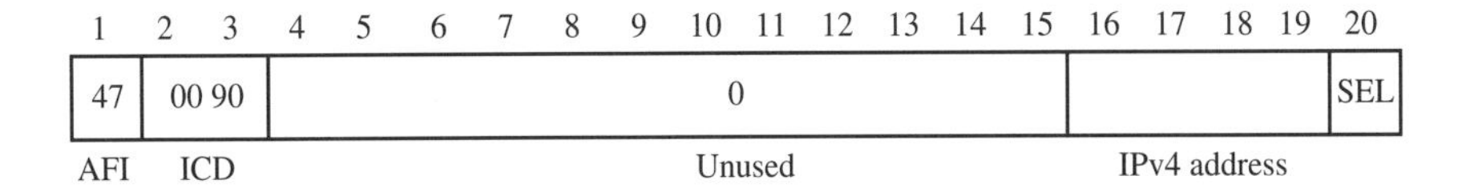

Figure 10.9 The NSAP ATM format for IP addresses.

NCANet is a production network in the state of North Carolina, used for research and education purposes. The ICD format, shown in Figure 10.8, was selected. The US GOSIP coded HO-DSP field was used, which consists of a 1-byte Domain Format Identifier (DFI) field, a 3-byte Administrative Authority (AA) Field, a 2-byte reserved field, a 2-byte Routing Domain Number (RDN) field, and a 2-byte AREA field. The fields were populated as follows (in hexadecimal):

- *AFI* = 47, indicates that an ICD ATM format is used.
- *ICD* = 0005, indicates that a GOSIP (NIST) coded HO-DSP field is used.
- *DFI* = 80, indicates that the next three bytes represent the administrative authority, in this case the Micro Electronics Center of North Carolina (MCNC), which is responsible for handling the regional traffic.
- *AA* = FFEA00, assigned to MCNC by GOSIP (NIST).
- *Reserved field* = 0000.
- *RDN* = xxxx, to be assigned by MCNC. For instance, North Carolina State University is part of NCANet, and it has been assigned the RND value of 0101.
- *AREA* = yyyy, to be assigned by the RDN owner. For instance, a group of ATM addresses in North Carolina State University have been assigned the AREA value of 1114.

As a result, all ATM addresses of ATM end devices and ATM switches in NCANet have the following NSAP prefix (in hexadecimal): 47.0005.80.FFEA00.0000.xxxx.yyyy.

The following address (in hexadecimal) is an example of the complete ATM address of an ATM switch in the ATM Lab of North Carolina State University:

47.0005.80.FFEA00.0000.0101.1114.400000000223.00.

The first 13 bytes is the prefix, and it is equal to 47.0005.80.FFEA00.0000.0101.1114. The next field is the ESI, and it is populated with the value of 400000000223, which is the IEEE MAC address of the switch. The final field is the selector (SEL), and it is populated with the value 00.

10.6 THE FORMAT OF THE SIGNALING MESSAGE

The format of the signaling message is shown in Figure 10.10. This message format is used by different signaling protocols, such as Q.2931, Q.2971 and PNNI. The protocol discriminator field is used to identify the signaling protocol. Bytes 3 to 5 give the call reference number to which the signaling message pertains. This is simply a number assigned to each call (i.e. connection) by the side that originates the call. It is a unique number that has local significance, and it remains fixed for the lifetime of the call. After the call ends, the call reference value is released, and it can be used for another call.

byte	1	2	3	4	5	6	7	8
byte 1	Protocol discriminator							
2	0	0	0	0	length of call ref. value			
3	Flag	Call reference value						
4	Call reference value							
5	Call reference value							
6	Message type							
7	Message type							
8	Message length							
9	Message length							
≥10	variable length information elements							

Figure 10.10 The signaling message format.

The call reference value is used by the signaling protocol to associate messages to a specific call, and it has nothing to do with the VPI/VCI values that will be assigned to the resulting ATM connection. The length of the call reference value is indicated in byte 2. For instance, 0011 indicates a 3-byte length. Since the call reference value is selected by the side which originates the call, it is possible that two calls originating at the opposite sides of the interface may have the same call reference value. The call reference flag, in byte 3, is used to address this problem. Specifically, the side that originates the call sets the flag to 0 in its message, whereas the destination sets the flag to 1 when it replies to a message sent by the originating side.

The message type field of the signaling message, bytes 6 and 7, is used to identify the type of the signaling message.

The message length field, bytes 8 and 9, is used to indicate the length of the signaling message, excluding the first nine bytes. Typically, there is a variety of information that has to be provided with each signaling message. This information is organized into different groups, known as *Information Elements* (IE). Each signaling message contains a variable

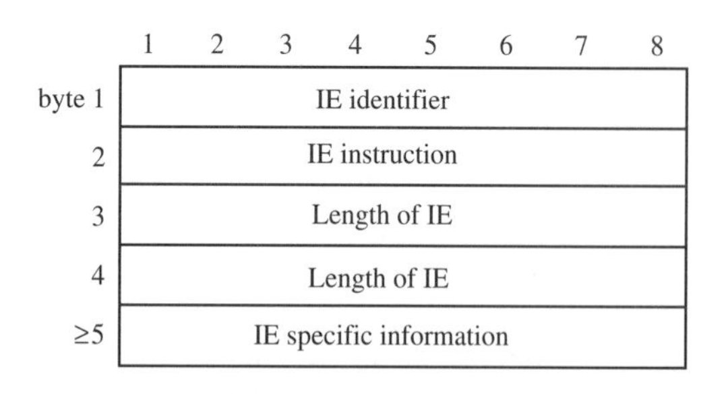

Figure 10.11 The structure of an information element.

number of information elements, which are appended to the signaling message starting at byte 10. The total length of all the information elements appended to a signaling message is given in the message length field. The structure of an information element is shown in Figure 10.11. The first byte contains the IE identifier, which is used to uniquely identify the information element. The second byte contains various fields, such as the coding standard, i.e. ITU-T, ISO/IEC, national, network specific standard (private or public), and the IE action indicator. Bytes 3 and 4 give the length of the information element, excluding the first four bytes, and the remaining bytes starting at byte 5 contain the information specific to the IE.

10.7 THE SIGNALING PROTOCOL Q.2931

We recall that this protocol is used to establish a point-to-point SVC over the private UNI in real-time. In this section, we first examine the information elements used in the Q.2931 messages, and then describe the Q.2931 messages, showing how they are used to establish and terminate a call. We will make use of the term *calling user or party* and *called user or party*. The calling user is a user in the end device that initiates a call, whereas the called user is the user at the end device that is being called.

10.7.1 Information Elements (IE)

Each signaling message contains a variety of information organized into different groups, known as Information Elements (IE). The following are some of the information elements used in Q.2931 messages:

- *AAL parameter IE*: used to indicate the AAL parameter values used between end devices.
- *ATM traffic descriptor IE*: used to specify the traffic parameters in the forward and backward direction of the connection.
- *Broadband bearer capability IE*: used to define the ATM service requested for a new connection.
- *Broadband high-layer IE, broadband low-layer IE*: used for compatibility checking by the called user.
- *Broadband repeat indicator IE*: used to indicate how repeated IEs are to be interpreted.
- *Call state*: used to describe the current status of the call.
- *Called party number IE, and called party sub-address IE*: used to identify the called user.
- *Calling party number IE, and calling party sub-address IE*: used to identify the calling user.
- *Cause IE*: used to describe the reason for generating certain messages, and indicates the location of the cause originator.
- *Connection identifier IE*: used to identify the VPI/VCI allocated to the connection at the UNI.
- *End-to-end transit delay IE*: used to indicate the maximum acceptable transit delay and the cumulative transit delay to be expected for the connection.
- *Extended QoS parameters IE*: used to specify the acceptable values and cumulative values of some of the QoS parameters.
- *Transit network selection IE*: used to identify a transit network that the call may cross.

An ATM end device or an ATM switch may not be able to process every information element included in a signaling message. In this case, the ATM equipment simply uses only the information elements that it needs, and it ignores the rest of them.

Below, we describe some of the information elements. For a detailed description of all the information elements and their fields, the reader is referred to the ATM Forum UNI signaling specification.

AAL IE

The purpose of this IE is to indicate the requested AAL and associated parameters. Some of the fields defined in this IE are: AAL type, bit rate (for AAL 1), source clock frequency recovery method, error correction method, structured data transfer block size, partially filled out cell method, forward and backward CPCS PDU size, MID range, and SSCS type.

ATM Traffic Descriptor IE

This IE is used to specify the traffic parameters in the forward and backward directions of the connection. We recall that a point-to-point connection is bidirectional. The traffic parameters and quality of service is defined separately for each direction.

In this IE, the PCR, SCR and MBS for the forward and backward directions can be specified as follows: forward PCR_0, backward PCR_0, forward PCR_{0+1}, backward PCR_{0+1}, forward SCR_0, backward SCR_0, forward SCR_{0+1}, backward SCR_{0+1}, forward MBS_0, backward MBS_0, forward MBS_{0+1}, and backward MBS_{0+1}. The subscript 0 implies that the traffic parameter applies to the traffic consisting of untagged cells, i.e. cells with $CLP = 0$. The subscript $0 + 1$ applies to traffic consisting of both tagged and untagged cells, i.e. cells with $CLP = 0$ or $CLP = 1$.

Other traffic parameters are also specified in this IE, such as best effort indicator used for UBR traffic, forward discard enable, backward discard enable, forward violation tagging allowed/disallowed, and backward violation tagging allowed/disallowed.

Additional network-specific code can be used to specify nonstandardized traffic parameters, such as average cell rate and average burst size.

Broadband Bearer Capability IE

This IE is used to indicate the requested bearer service. It contains the fields: bearer class, and *ATM Transfer Capability* (ATC).

The following five bearer classes have been defined:

- Broadband connection oriented bearer service, class A (BCOB-A),
- Broadband connection oriented bearer service, class C (BCOB-C),
- Frame relay bearer service,
- Broadband connection oriented bearer service, class X (BCOB-X), and
- Transparent VP service.

BCOB-A is a connection-oriented service that provides a constant bit-rate and end-to-end timing requirements. BCOB-C is also a connection-oriented, variable-bit rate service with no end-to-end timing requirements. BCOB-X is a connection-oriented service where the AAL, traffic type (VBR or CBR) and timing requirements are defined by the user. The frame relay bearer service, as its name implies, is for frame relay services. Finally, in the transparent VP service, both the VCI field and the payload type indicator field are transported transparently by the network. We recall that in Section 5.1 of Chapter 5, we introduced the service classes A, B, C, D and X. BCOB-A is the same as class A, BCOB-C is the same as class C, and BCOB-X is the same as class X.

The ATM Transfer Capability (ATC) field gives the ATM service class requested by the calling party. We recall from Section 7.3 of Chapter 7 that an ATM transfer capability is ITU'T's term for an ATM service category. The following service categories can be defined in the IE:

- CBR,
- CBR with CLR commitment on the traffic with $CLP = 0 + 1$,
- Real-time VBR,
- Real-time VBR with CLR commitment on the traffic with $CLP = 0 + 1$
- Non-real-time VBR,
- Non-real-time VBR with CLR commitment on the traffic with $CLP = 0 + 1$,
- ABR,
- ATM block transfer-delayed transmission, and
- ATM block transfer-immediate transmission.

End-to-End Transit Delay IE

This IE is used to indicate the maximum acceptable transit delay and the cumulative transit delay to be expected for the connection. These delays are expressed in msec.

The transit delay is defined as the one-way, end-to-end delay between the calling user and the called user. It consists of the total processing delay in the end device of the calling user, the total processing delay in the end device of the called user, and the max CTD (see Section 7.2). The total processing delay in either end device consists of the AAL handling delay, the ATM cell assembly delay, and other processing delays.

The maximum acceptable transit delay may be indicated by the calling user in its SETUP message to the network. The cumulative transit delay indicated in the SETUP message by the calling user includes only the total processing delay at its end device. The network indicates the end device total processing delay, plus the network transfer delay to the end device of the called user. The network transfer delay is calculated cumulatively by adding up the propagation delay on each hop and the switching delay in each ATM switch that lies on the path of the connection. The cumulative transit delay, which includes the total processing delays in both end devices and the network transfer delay, is communicated in the CONNECT message.

Extended QoS Parameters IE

The information specified in this IE is in addition to the information specified in the end-to-end transit delay IE. It is used to indicate the values of the peak-to-peak cell delay variation, and the CLR (see Section 7.2). The acceptable value for each of these two QoS parameters can be specified in the forward and backward directions. Also, the cumulative values for each of these two QoS parameters can be indicated in the forward and backward direction.

10.7.2 Q.2931 messages

The Q.2931 messages can be grouped into the following three categories: call establishment, call clearing and miscellaneous. The messages for each category are given in Table 10.2. Each Q.2931 message uses the signaling message format described in Section 10.6, with the protocol discriminator set to 00001001, and it contains a set of information elements. Below, we describe the function of each message, and then show how they are used to establish and clear a call.

- *ALERTING*: this message is sent by the called user to the network and by the network to the calling user to indicate that 'called user alerting' has been initiated. Called user alerting is used for calls that require human interface, such as voice. The information element used in this message is the connection identifier.
- *CALL PROCEEDING*: the message is sent by the called user to the network, or by the network to the calling user, to indicate that the requested establishment of the call has been initiated and no more call information is accepted. The information element used in this message is the connection identifier.
- *CONNECT*: the message is sent by the called user to the network, or by the network to the calling user, to indicate that the called user has accepted the call. The following information element are used in this message: AAL parameter, broadband low-layer, connection identifier, and end-to-end transit delay.
- *CONNECT ACKNOWLEDGMENT*: this message is sent by the network to the called user to indicate that the user has been awarded the call. It is also sent by the calling user to the network to allow symmetrical call control procedures.
- *RELEASE*: this message is sent by the user to request the network to clear an end-to-end connection. It is also sent by the network to indicate that an end-to-end connection is cleared, and that the receiving equipment should release the connection identifier and prepare to release the call reference value after sending RELEASE COMPLETE. The cause information element is carried in this message.
- *RELEASE COMPLETE*: this message is sent by the calling user or the network to indicate that the equipment sending the message has released its call reference value and, if appropriate, the connection identifier. The cause information element is carried in this message.
- *SETUP*: this message is sent by the calling user to the network and by the network to the called user to initiate the establishment of a new call. The following information elements are used: AAL parameter, ATM traffic descriptor, broadband bearer capability,

Table 10.2 The Q.2931 messages.

Call establishment messages	ALERTING, CALL PROCEEDING, CONNECT, CONNECT ACKNOWLEDGMENT, SETUP
Call clearing messages	RELEASE, RELEASE COMPLETE
Miscellaneous messages	NOTIFY, STATUS, STATUS ENQUIRY

broadband high-layer, broadband low-layer, called party number, called party subaddress, calling party number, calling party subaddress, connection identifier, end-to-end transit delay, extended QoS parameters, and transit network selection.

The NOTIFY message is sent by the user or the network to indicate information pertaining to a call. The STATUS message is sent by the user or the network in response to a STATUS ENQUIRY message. Finally, the STATUS ENQUIRY message is sent by the user or the network to solicit a STATUS message from the peer Q.2931 protocol.

Call Establishment

The steps involved in establishing a call are shown in Figure 10.12. The calling user initiates the procedure for establishing a new call by sending a SETUP message to its ingress ATM switch across the UNI. The ingress switch sends a CALL PROCEEDING message to the calling user if it determines that it can accommodate the new call. (If it cannot accommodate the new call, it rejects it by responding with a RELEASE COMPLETE message.) The ingress switch calculates a route to the destination end device over which the signaling messages are transferred. The same route is used to set up a connection over which the data will flow. It then forwards the SETUP message to the next switch on the route. The switch verifies that it can accommodate the new connection, and then forwards the SETUP message to the next switch, and so on, until it reaches the end device of the called user. The PNNI protocol, described in Chapter 11, is used to progress the SETUP message across the network.

If the called user can accept the call, it responds with a CALL PROCEEDING, ALERTING or CONNECT message. (Otherwise, it sends a RELEASE COMPLETE message.) Upon receiving an indication from the network that the call has been accepted, the ingress switch sends a CONNECT message to the calling user, which responds with a CONNECT ACKNOWLEDGMENT.

Call Clearing

Call clearing is initiated when the user sends a RELEASE message. When the network receives the RELEASE message, it initiates procedures for clearing the connection to the

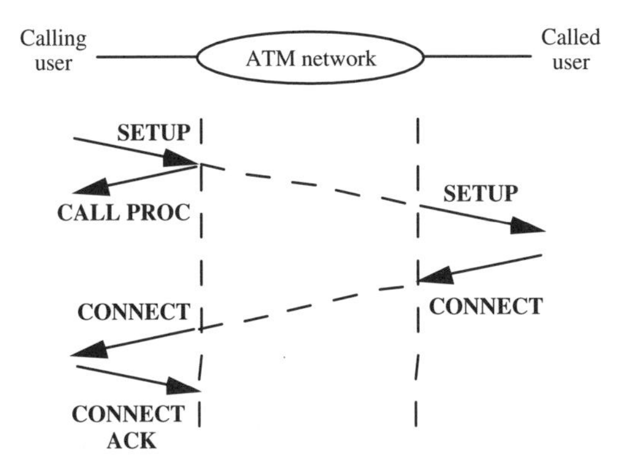

Figure 10.12 Call establishment.

remote user. Once the connection has been disconnected, the network sends a RELEASE COMPLETE message to the user, and releases both the call reference value and the connection identifier. Upon receipt of RELEASE COMPLETE message, the user releases the connection identifier and the call reference value.

10.8 THE SIGNALING PROTOCOL Q.2971

A point-to-multipoint connection is always unidirectional, and it allows an end device to send its traffic to two or more end devices. The end device that generates the traffic is known as the *root*, and the receiving end devices are known as the *leaves*.

To establish a point-to-multipoint connection over the private UNI, the root first establishes a point-to-point connection to a single leaf using the Q.2931 signaling procedures. The Q.2971 signaling procedures are used then to add or drop new leaves.

The Q.2971 messages utilize the information elements described in Section 10.7.1, and two new information elements, namely the *endpoint reference* IE, and the *endpoint state* IE. The purpose of the endpoint reference IE is to identify an individual leaf of in a point-to-multipoint connection. The endpoint state IE is used to indicate the state of a leaf in a point-to-multipoint connection. The following states have been defined: null, add party initiated, party alerting delivered, add party received, party alerting received, drop party initiated, drop party received and active.

The following messages are used by Q.2971: ADD PARTY, ADD PARTY ACKNOWLEDGMENT, PARTY ALERTING, ADD PARTY REJECT, DROP PARTY, DROP PARTY ACKNOWLEDGMENT. Below, we discuss the function of each message, and give examples of how they are used to add or drop a leaf.

- *ADD PARTY*: this message is used to add a new leaf to an already established point-to-multipoint connection. It is sent from the root to the network to request the addition of a new leaf. The following information elements are used: AAL parameter, broadband high-layer, broadband low-layer, called part number, called party subaddress, calling party number, calling party subaddress, end-to-end transit delay, transit network selection, and the endpoint reference IE.
- *ADD PARTY ACKNOWLEDGMENT*: this message is sent from the network to the root to indicate that the ADD PARTY message was successful. The AAL parameter, broadband low-layer, and end-to-end transit information elements are used in this message.
- *PARTY ALERTING*: this message is sent from the network to the root to notify it that alerting of the called party has been initiated. The endpoint reference IE is used.
- *ADD PARTY REJECT*: this message is sent from the network to the root to indicate that the ADD PARTY request was not successful. The endpoint reference IE and the cause IE are used in this message.
- *DROP PARTY*: this message is sent either from the root to the network or from the network to the root to request to drop a leaf from the connection. The endpoint reference IE is used in this message.
- *DROP PARTY ACKNOWLEDGMENT*: this message is sent from the root to the network or from the network to the root to acknowledge that the connection to a leaf has been successfully cleared. The endpoint reference IE and the cause IE are used in this message.

Establishment of a Point-to-Multipoint Connection

The connection to the first leaf is set up using Q.2931, following the procedures described in Section 10.7.2. The SETUP message sent by the root is required to contain an endpoint reference IE and the broadband bearer capability IE with the indication that the connection is a point-to-multipoint. The traffic parameters in the traffic descriptor IE are only defined in the forward direction, since the connection is unidirectional. The traffic parameters for the backward direction are set to zero.

Adding a new leaf to the connection is effected by sending an ADD PARTY message, as shown in Figure 10.13. Each ADD PARTY message has the same call reference value as that specified in the initial SETUP message used to set up the point-to-multipoint connection to the first leaf. The network issues a SETUP message to the called party, which responds with a CALL PROCEEDING message, and then with a CONNECT message. The calling party finally receives an ADD PARTY ACKNOWLEDGMENT, which confirms the addition of the new leaf.

Dropping a Leaf

A leaf can be dropped from a point-to-multipoint connection due to a request by the root, or by the leaf itself. When a leaf wants to drop from the point-to-multipoint connection,

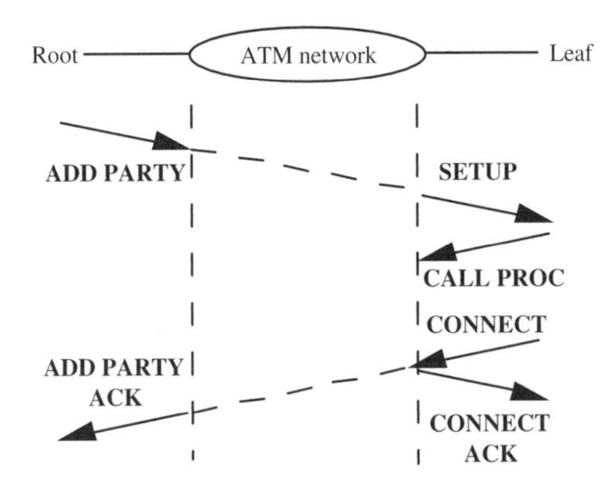

Figure 10.13 Adding a leaf to the point-to-multipoint connection.

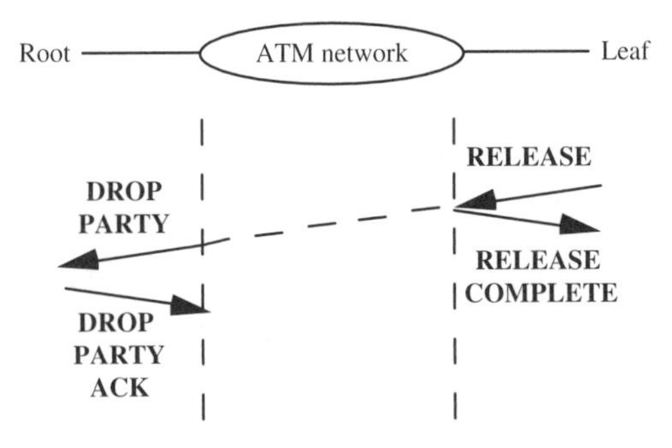

Figure 10.14 Dropping a leaf.

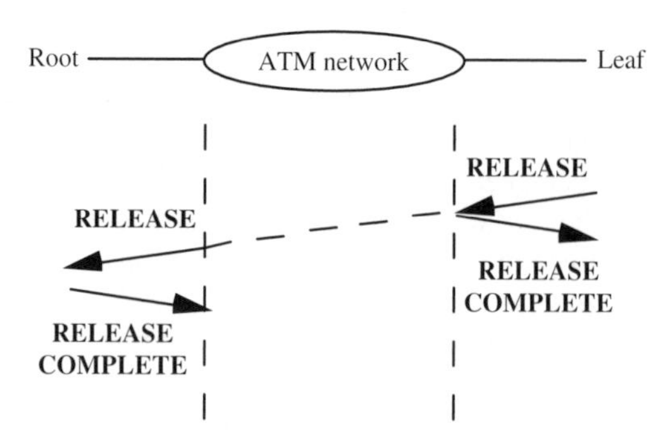

Figure 10.15 Dropping the last leaf.

it sends a RELEASE message to the network, as shown in Figure 10.14. This causes the network to send a DROP PARTY message to the root, which responds with a DROP PARTY ACKNOWLEDGMENT message.

When the root wants to drop a leaf, it issues a DROP PARTY message to the network. This causes the network to send a DROP PARTY message to the leaf, to which the leaf responds with a DROP PARTY ACKNOWLEDGE message. The DROP PARTY ACKNOWLEDGE message is delivered by the network to the root.

When the last leaf is dropped from a point-to-multipoint connection, the call is released as shown in Figure 10.15.

10.9 LEAF INITIATED JOIN (LIJ) CAPABILITY

The Q.2971 procedures described in the previous section allow the root of a connection to add a leaf to its point-to-multipoint connection. It is not possible using Q.2971 for a leaf to join a point-to-multipoint connection without the intervention from the root of the connection. This can be achieved using ATM Forum's *Leaf Initiated Join* (LIJ) capability. Two modes of operation have been defined for the LIJ capability, *leaf-prompted join without root notification* and *root-prompted join*.

In the leaf-prompted join without root notification mode, a leaf can send a request over its UNI to join a particular point-to-multipoint connection. If the leaf's request is for an existing connection, then the request is handled by the network, and the root is not notified when the leaf is added or dropped from the connection. If the leaf's request is for a connection that is not yet established, then the request is forwarded to the root, which then performs the initial set-up of the connection. This type of connection, where the network supports the joining and leaving of leaves is referred to as a *network LIJ connection.*

In the root-prompted join mode of operation, a leaf can send a request over its UNI to join a point-to-multipoint connection, but the request is handled by the root. The root adds and drops leaves from a connection following the Q.2971 procedures described above. This type of connection is called the *root LIJ connection.*

The root LIJ connection should not be confused with the connection that can be set-up using Q.2971. For, in Q.2971, it is the root that initiates a join in order for an end device to be attached to the point-to-multipoint connection as a leaf. It is beyond the scope of the signaling protocol to provide the root with the ATM addresses of the end devices that wants to join the connection. A good example of how the root gets this information is in IP multicasting over ATM, described in Section 8.3.2. In this case, we saw that MARS keeps a record of all the ATM addresses of the end devices that want to be part of an IP multicast address. MARS knows about these end devices, because each end device is required to register with MARS. The set of all ATM addresses associated with an IP multicast is known as the *host map*. MARS downloads the host map to the root (in the VC mesh case) or to the multicast server (in the multicast server case), which is responsible for setting up, maintaining and releasing the point-to-multipoint connection. As we can see, in this case the root is simply provided with the ATM addresses of all the leaves.

To support the LIJ capability, two new signaling messages were defined, namely LEAF SETUP REQUEST and LEAF SETUP FAILURE. The LEAF SETUP REQUEST is used by an end device to request to join a point-to-multipoint connection. If this request fails, the network sends back to the end device a LEAF SETUP FAILURE message.

Also, the following three new information elements were defined:

- *Leaf Initiated Join (LIJ) call identifier IE*: used to uniquely identify a point-to-multipoint call at a root's interface. The LIJ call identifier is specified in the SETUP message when the root creates a point-to-multipoint call with LIJ capability. It is also specified in the LEAF SETUP REQUEST message when a leaf wishes to join the identified call. How a leaf knows a LIJ call identifier is outside the scope of this specification. (It could, for instance, be obtained through a directory service.)
- *Leaf Initiated Join (LIJ) parameters IE*: used by the root to associate options with the call when the call is created. Specifically, it can indicate whether the connection is a network LIJ connection.
- *Leaf sequence number IE*: used by a joining leaf to associate subsequent signaling messages triggered by a LEAF SETUP REQUEST message that it sent over its UNI.

When a leaf wishes to join a network or root LIJ connection, it sends a LEAF SETUP REQUEST message across its UNI. This message includes the following IEs: calling party number, calling party subaddress, called party number, called party subaddress, LIJ call identifier, leaf sequence number, and transit network selection. The leaf sequence number is used to associate signaling messages sent to the leaf in response to this particular LEAF SETUP REQUEST message. The LIJ call identifier gives the identifier of the connection that the leaf wants to join. The called party number is the address of the root associated with this particular LIJ call, and the calling party number is the leaf's address. As an option, the leaf may include the transit network selection IE. The network uses this information to route the LEAF SETUP REQUEST message towards the root.

If the LEAF SETUP REQUEST is for a network LIJ connection, and the connection already exists, the network sends a SETUP or ADD PARTY message to the leaf, as shown in Figure 10.16. If the LEAF SETUP REQUEST is for a root LIJ connection, and the connection already exists, the network forwards the message to the root. To add the leaf to the connection, the root issues an ADD PARTY message following the Q.2971 procedures, as shown in Figure 10.17.

If the connection does not exist, but for which a valid root has been specified, the LEAF SETUP REQUEST message will be delivered to the root by the network. Upon receipt, the root creates a point-to-multipoint call using Q.2971. The root has the choice of creating either a network LIJ call or a root LIJ call.

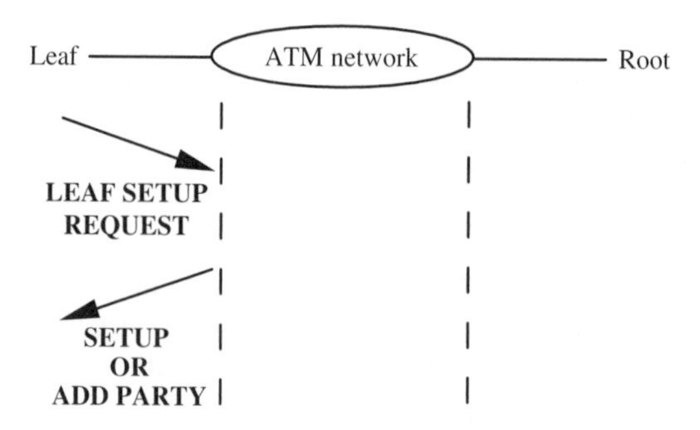

Figure 10.16 Joining an existing network LIJ connection.

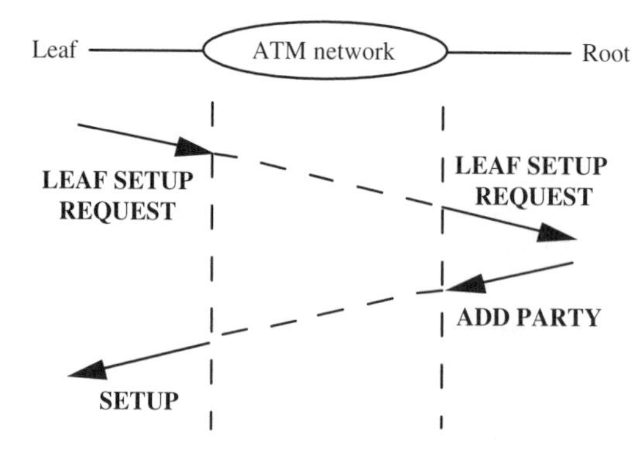

Figure 10.17 Joining an existing root LIJ connection.

Finally, if the network is unable to complete the LEAF SETUP REQUEST message for any reason, it sends a LEAFSETUP FAILURE message back to the leaf.

10.10 ATM ANYCAST CAPABILITY

This capability allows a user to request a point-to-point connection to a single end device that is part of an ATM group address. An ATM group address is a collection of one or more ATM end devices. An ATM end device can be a member of zero or more ATM groups at any time. An end device that is a member of an ATM group has two types of addresses: an individual address and a group address.

The ATM anycast capability can be requested by a calling user by sending a SETUP message across its UNI. The SETUP message contains the desired ATM group address in the called party number IE. When the connection request reaches a group member, the called member may return its own individual ATM address.

A well known group address is used to identify an ATM group that implements a well known service. An example of such a service is that provided by a LAN emulation configuration server. This server is used to assign an individual LE client to a particular

LAN emulation server. ATM Forum well-known addresses are assigned for use in ATM Forum specifications.

PROBLEMS

1. Why does the selective-reject ARQ described in Section 2.3 not work well in a network with a high bandwidth-delay product?
2. What are the basic differences between the error recovery scheme in the SSCOP and the more traditional ARQ schemes, such as go-back-n and selective reject?
3. Describe the sequence of primitives issued to set up a connection between two peer signaling protocols.
4. Identify the subfields of the NSAP address of an ATM switch in your organization.
5. What is the purpose of the call reference flag in the signaling message?
6. In which information element does the calling user indicate its traffic parameters?
7. In which information elements does the calling user indicate the quality-of-service parameters?
8. Trace the sequence of the signalling messages issued to set up a connection.
9. Trace the sequence of the signalling messages issued to add a leaf to a point-to-multipoint connection.
10. What is the main difference between Q.2971 and the LIJ capability?

11

The Private Network-Network Interface (PNNI)

In the previous chapter, we examined the signaling procedures used to set up a Switched Virtual Connection (SVC) across the private UNI in real-time. In this chapter, we examine the *Private Network-Network Interface*, or *Private Network Node Interface* (PNNI) protocol, used to establish an SVC across a private network. The PNNI protocol consists of the *PNNI routing protocol* and the *PNNI signaling protocol*. The PNNI routing protocol is used to distribute network topology and reachability information between the switches of a private network. The PNNI signaling protocol is used to establish, maintain and clear ATM connections in a private network. We first describe the PNNI routing protocol in detail, and then briefly discuss the PNNI signaling protocol.

11.1 INTRODUCTION

Prior to the development of the PNNI protocol, the only standardized mechanism for ATM routing and signaling was the *Interim Interswitch Signaling Protocol* (IISP). This protocol uses manually configured static routes and UNI signaling between switches to forward signaling requests across an ATM network. IISP was defined as an interim step until the ATM Forum completed phase I of the PNNI protocol.

As shown in Figure 11.1, PNNI provides the interface between two ATM switches that either belong to the same private ATM network or to two different private ATM networks. The abbreviation PNNI can be interpreted as either the *private network node interface* or the *private network-network interface*, reflecting these two possible uses.

The PNNI protocol consists of two components, namely, the *PNNI routing protocol* and the *PNNI signaling protocol*. The PNNI routing protocol is used to distribute network topology and reachability information between switches and clusters of switches. This information is used to compute a path from the ingress switch of the source end device to the egress switch of the destination end device over which signaling messages are transferred. The same path is used to set up a connection along which the data will flow. PNNI was designed to scale across all sizes of ATM networks, from a small campus network with a handful of switches, to large worldwide ATM networks. Scalability is achieved by constructing a multi-level routing hierarchy based on the 20-byte ATM NSAP addresses. As will be seen, this level of scalability requires a significant complexity in the PNNI protocol.

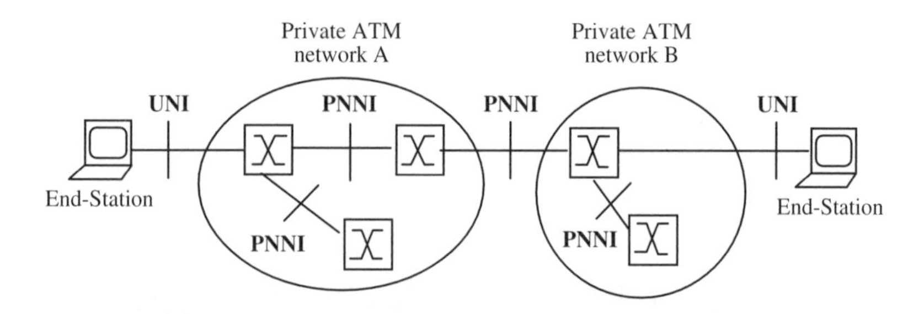

Figure 11.1 The private network-network interface.

The PNNI signaling protocol is used to dynamically establish, maintain and clear ATM connections at the private network-network interface and at the private network node interface.

11.2 THE PNNI ROUTING PROTOCOL

Let us consider the network of private ATM switches shown in Figure 11.2. Each circle represents an ATM switch, and a straight line connecting two circles indicates a physical link. The addresses of the nodes used in this example are fictitious, and they are purely for presentation purposes. In real-life, each node is assigned a 20-byte ATM NSAP address, described in Section 10.5. We assume that various end devices, not shown in Figure 11.2, are attached to these switches. Each end device also has an ATM NSAP address. We recall that only the first 19 bytes of the NSAP address are used to identify an end device. The last 1-byte selector (SEL) field has only local significance to the end device, and it is used to distinguish different destinations reachable at the end device.

Now, let us assume that an end device issues a SETUP message to its ingress switch. The switch calculates a path to the egress switch of the destination end device, and then it forwards the SETUP message to the next switch along the path. For the switch to calculate the path to the destination egress switch, it has to know the topology of the network, and also it has to know the ATM addresses of the end devices attached to each switch. If the network is small, then it is feasible for each switch to have the complete network topology and reachability information. However, it becomes infeasible when dealing with large networks. PNNI addresses this issue by organizing the network in a hierarchical structure, designed to reduce the amount of topology and reachability information a switch has to keep in order to calculate a path to a destination egress switch. Below, we describe how PNNI works using the example shown in Figure 11.2, which was taken from the ATM Forum's PNNI standard.

11.2.1 The Lowest-Level Peer Groups

The PNNI hierarchy starts at the lowest level, which consists of the actual ATM switches. These switches are referred to as the *lowest-level nodes*. These lowest-level nodes are organized into *Peer Groups* (PG). The peer groups are then organized into higher-level peer groups, and so on until a hierarchical structure is constructed which encompasses the entire network.

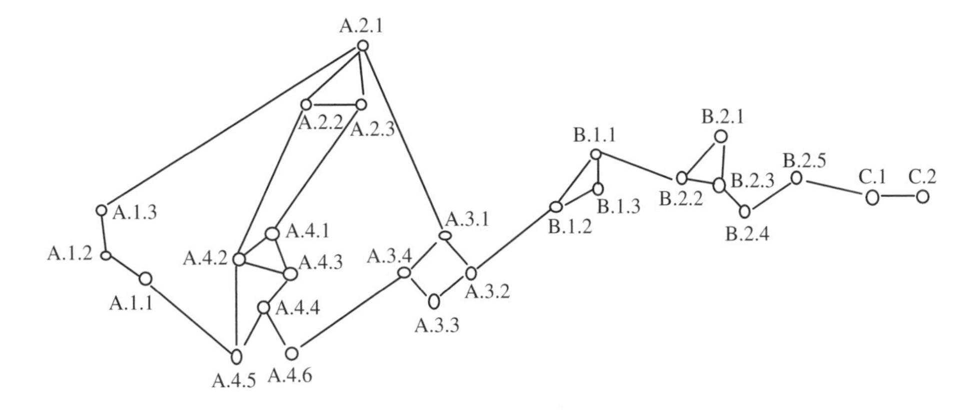

Figure 11.2 A network of ATM switches.

As shown in Figure 11.3, the lowest-level nodes are organized into peer groups A.1, A.2, A.3, A.4, B.1, B.2 and C. Peer group A.1, referred to as PG(A.1), consists of the lowest-level nodes A.1.1, A.1.2 and A.1.3, peer group A.2, referred to as PG(A.2), consists of the lowest-level nodes A.2.1, A.2.2, A.2.3, etc.

The organization of a set of lowest-level nodes into a peer group is done administratively, by configuring each lowest-level node with the same *peer group identifier*. A peer group identifier is a common prefix of the ATM NSAP addresses of the lowest-level nodes, and it can be at most 13 bytes long (see Section 10.5). We recall that the first 13 bytes of an ATM NSAP address correspond to the fields IDP and HO-DSP. The length of this prefix, that is the number of bits that it consists of, is known as the *level indicator*. Since the longest prefix can be 13 bytes, the longest level indicator can be 104. Neighboring nodes exchange peer group identifiers in *hello* packets using the *hello* protocol. If they find that they have the same peer group identifier, then they belong to the same peer group.

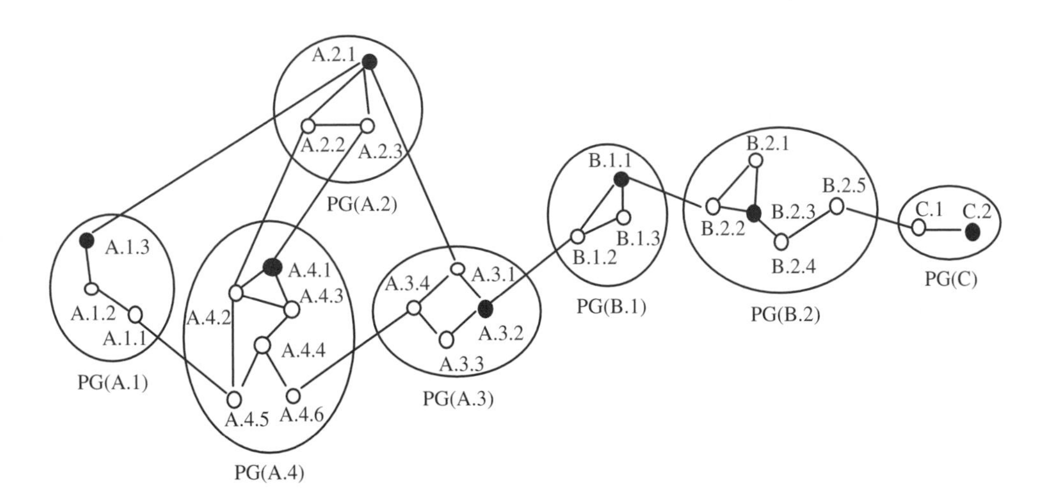

Figure 11.3 Peer groups of the lowest-level nodes.

The lowest-level nodes in a peer group are connected via physical links. When a physical link becomes operational, the two nodes on either side of the link initiate an exchange of information through hello packets. Each node announces the ATM addresses of the end devices attached to it, the node's ATM address, and the port identifier of the physical link. Each node then bundles this information into *PNNI Topology State Elements* (PTSE), and floods them reliably throughout the peer group. Each node also generates a PTSE that describes its own identity and capabilities. Reliable flooding is done as follows. The PTSEs are carried in *PNNI Topology State Packets* (PTSP). When a PTSP is received, it is acknowledged and then sent to all the other neighbors of the node. The PTSEs are subject to aging, and they get removed after a predefined duration, unless they are refreshed. So, PTSEs are issued periodically, and also when a new link becomes operational. As a result of this mechanism, each node in the peer group has an identical copy of the topology database.

It is possible that a node in a peer group may have one or more links to other nodes, which do not belong to the same peer group. In this case, this node is referred to as a *border* node. For instance, node A.1.1 is connected to A.4.5 via a physical link, and in view of this, it is a border node as far as peer group A.1 is concerned. Likewise, A.4.5 is a border node as far as peer group A.4 is concerned. As we will see, border nodes are important in the creation of the PNNI hierarchy.

11.2.2 The Next Level of Peer Groups

The peer groups themselves get organized into higher-level peer groups, as shown in Figure 11.4. In this example, the new peer groups are A and B, referred to in the figure

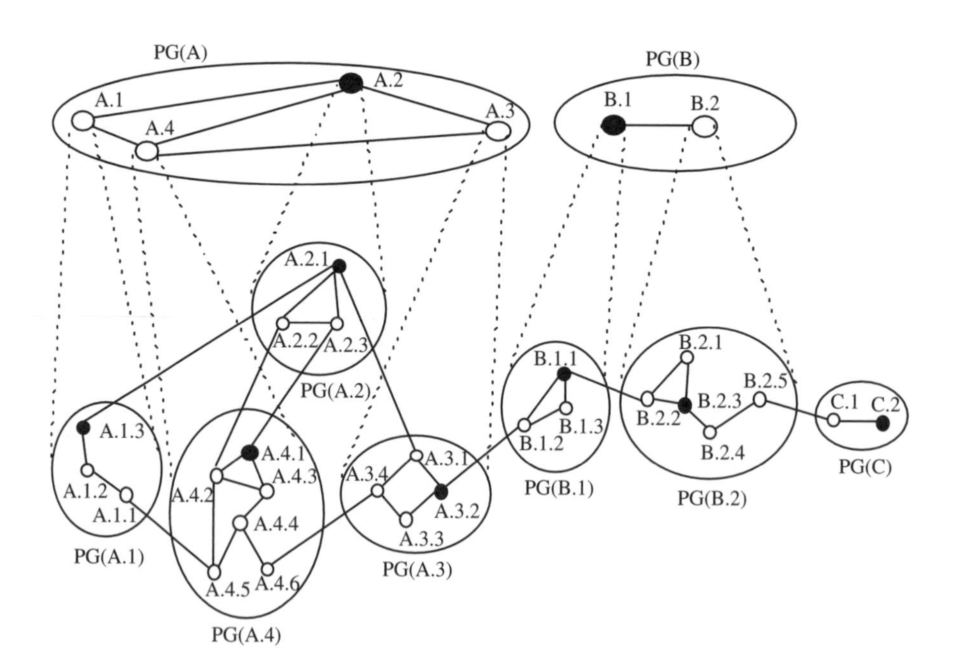

Figure 11.4 Second level of peer groups.

as PG(A) and PG(B). Peer group A consists of the lower-level peer groups A.1, A.2, A.3 and A.4, and peer group B consists of the lower-level peer group B.1 and B.2.

Each of the lower-level peer groups is represented in the higher-level peer group as a single node, known as the *logical group node*. For instance, in peer group A, node A.1 is the logical group node that represents peer group A.1, node A.2 is the logical group node that represents peer group A.2, and so on. A logical group node is an abstraction of a peer group for the purpose of representing it in the next level of the PNNI hierarchy. Logical group nodes are connected by *logical links*, which may be either physical links or SVCs.

A logical group node is implemented in one of the lowest-level nodes of the peer group that it represents. It has an ATM address which is basically the address of the lowest-level node on which it has been implemented, but with a different selector (SEL) value. The same lowest-level node is also the *Peer Group Leader* (PGL). The peer group leader for each peer group in Figures 11.3 and 11.4 is indicated by a black circle. The peer group leader is elected from among the lowest-level nodes of its peer group as follows. Each lowest-level node in the peer group is configured with a leadership priority which is distributed to the nodes of its peer group. The node with the highest priority becomes the peer group leader.

The only function that a logical group node performs is to distribute aggregated and summarized information to its peer group about the peer group that it represents. A logical group node also passes information from its own peer group to the peer group leader of the peer group that it represents. The logical group node does not participate in the PNNI signaling. The peer group leader's function is to distribute information received by the logical group node to all the members of its peer group. Otherwise, the peer group leader acts like the other nodes in its peer group.

Peer group A is called the *parent peer group* of the peer groups A.1, A.2, A.3 and A.4, and each of the peer groups A.1, A.2, A.3 and A.4 is called the *child peer group* of peer group A.

11.2.3 Uplinks

A logical link connecting two logical group nodes may be a physical link or an SVC. PNNI links are classified into *horizontal, exterior* and *outside* links. A horizontal link connects two nodes within the same peer group. An exterior link connects a node within a peer group to a node which does not operate the PNNI protocol. Finally, an outside link connects two nodes within two different peer groups. For example, in Figure 11.4, the link that connects the lowest level nodes A.4.5 and A.4.4 is a horizontal link, whereas the link that connects the lowest level nodes A.1.1 and A.4.5 is an outside link. Two nodes connected with an outside link are in fact border nodes. This definition of links and border nodes holds at every level of the PNNI hierarchy.

Border nodes extend the hello protocol across outside links to include information about their respective higher-level peer groups and the logical group nodes representing them in these peer groups. This information allows the border nodes to determine that they have a common higher-level peer group, and also determine the higher-level logical group nodes to which the border nodes are connected. For instance, border node A.3.4 recognizes that its neighbor A.4.6 is represented by logical group node A.4. Consequently, A.3.4 advertises its link between itself and A.4. This link is known as an *uplink*, and node A.4

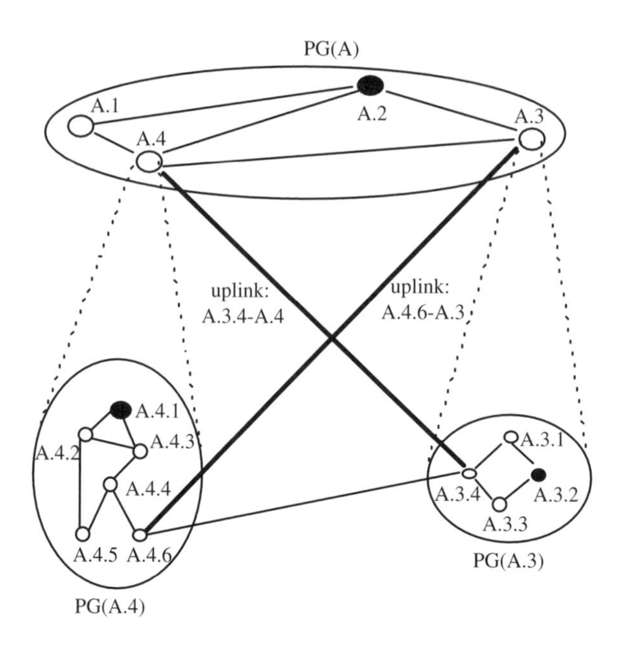

Figure 11.5 Uplinks and upnodes.

is known as an *upnode*. Similarly, node A.4.6 advertises its uplink to A.3. These two uplinks are shown in Figure 11.5. Border nodes advertise their uplinks in PTSEs flooded in their respective peer groups. This enables all the nodes in the peer group to update their topology database with these new uplinks. This information is also fed up to the logical group nodes by the peer group leaders. As a result, logical group nodes A.4 and A.3 will become aware that they belong to the same peer group, and they will establish a logical link, which will be an SVC.

11.2.4 Information Exchange in the PNNI Hierarchy

We recall that when a physical link is activated, the two attached lowest-level nodes initiate an exchange of PNNI related information. Similarly, when a logical link becomes operational, the attached logical group nodes initiate an exchange of information. Each node on the link sends hello packets to the node on the other side of the link, specifying the addresses of the end devices that are attached to it, its own ATM address, port identifier for the link, and link status information such as the total allocated bandwidth. Hello packets also support the exchange of peer group identifiers so that neighboring nodes can determine if they belong to the same peer group or different peer group.

The exchange of PNNI routing information over a physical link is done via the *Routing Control Channel* (RCC) designated with VPI = 0 and VCI = 18. The exchange of PNNI routing information between logical group nodes is done by an SVC. The VPI/VCI values for this SVC are assigned in the normal way when it is set-up. The PNNI routing protocol runs on top of AAL 5.

Feeding Information up the Hierarchy

Within each peer group, the peer group leader has a complete topology state information from all the nodes in its peer group. This information is fed to the logical group node that represents the entire peer group in the parent peer group, which in turn floods it to all the nodes in the parent peer group. The information consists of reachability and topological aggregation.

The reachability information consists of summarized addresses reachable through the lower level peer group. Address summarization is discussed in Section 11.2.7.

Topology aggregation refers to summarized topology information used to route a new connection across a peer group. This aggregation process reduces the amount of information that is needed to be exchanged, and consequently, it facilitates scalability in a large network. There are two types of aggregation: *link aggregation* and *nodal aggregation.* In link aggregation, a set of links between two peer groups is aggregated to a single logical link. For instance, in Figure 11.4, the two links connecting peer group A.2 and A.4 (i.e. links A.2.2–A.4.2 and A.2.3–A.4.1) are represented by a single link. In nodal aggregation, the topology of an entire peer group is represented in the parent peer group by a *complex* node, which indicates the aggregate links between the peer group and other peer groups.

Feeding Information down the Hierarchy

This is necessary to allow nodes in lower level peer groups to route to all destination reachable via the PNNI hierarchy. Each logical group node feeds down information to its peer group leader, which in turns floods it to all the nodes of the peer group. The information consists of all the PTSEs it originates or receives via flooding from the other members of its peer group. PTSEs flow horizontally among the nodes of a peer group, and downwards into and through child peer groups.

11.2.5 The Highest-Level Peer Group

Let us now go back to our example, and complete the construction of the PNNI hierarchy. We note in Figure 11.4 that we have not as yet obtained connectivity for all the lowest-level nodes. This can be achieved by configuring peer groups A, B and C into a single parent peer group, as shown in Figure 11.6. (Another possibility would be to create a parent peer group with nodes B and C, and then aggregate that with node A to form the highest level peer group.)

This is achieved following the same procedures as before. The logical group nodes in peer groups A and B elect a peer group leader, which then instantiates a logical group node in the higher-level peer group. Specifically, A.2 is the peer group leader of peer group A, B.1 is the peer group leader in peer group B, and C.1 is the peer group leader in peer group C. Each of these three peer group leaders instantiates a logical group node, which is a member of the higher-level peer group.

PNNI permits the construction of asymmetrical hierarchies. That is, a parent peer group can also be a grandparent or a great-grandparent peer group to some other lower level peer group. For example, the lower-level peer group C in Figure 11.6, is directly represented in the highest-level peer group by logical group node C, whereas lower-level peer groups B.1 and B.2 are first grouped into the parent peer group B before they are

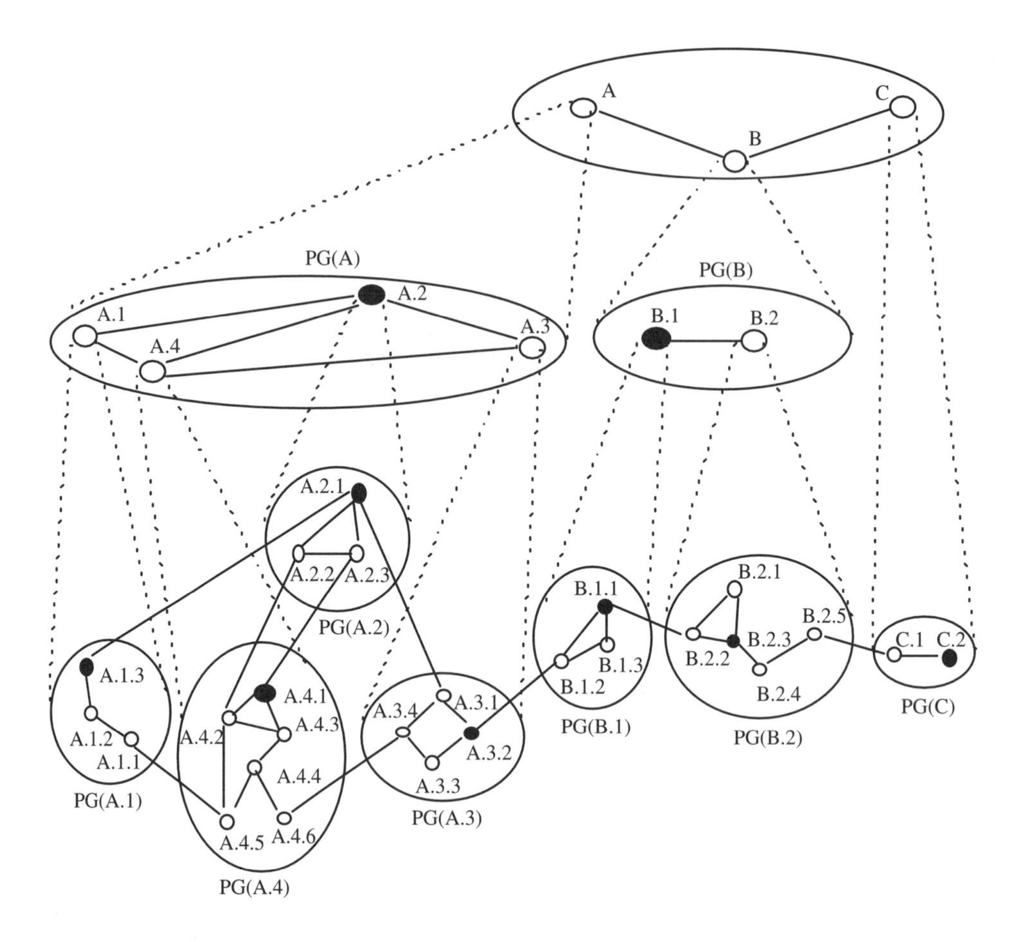

Figure 11.6 The complete PNNI hierarchy.

represented at the highest-level peer group by logical group node B. In view of this, the highest-level peer group is grandparent to peer groups B1 and B2, and a parent to peer group C.

The uplinks are shown in Figure 11.7. We can see that nodes C.1 and B.2.5 are border nodes connected with the outside link B.2.5–C.1. Consequently, following the discussion on uplinks in Section 11.2.3, C.1 advertises an uplink to B, and B.2.5 an uplink to C. Likewise, B.2.2 advertises an uplink to B.1 and B.1.1 advertises an uplink to B.2. The uplink B.2–C, is known as an *induced* uplink, and it is derived as follows. When the peer group leader B.2.3 receives the PTSE flooded by B.2.5 describing the uplink B.2.5–C, it passes the information to the logical group node B.2. This information consists of the common peer group identifier (the highest-level peer group, in this case) and the ATM address of the upnode C. From this information, B.2 recognizes that node C is not a member of peer group B, and it derives the new uplink B.2–C.

Nodes B and C, through the information they receive regarding the uplinks, recognize that they belong to the same peer group, and they establish an SVC between them.

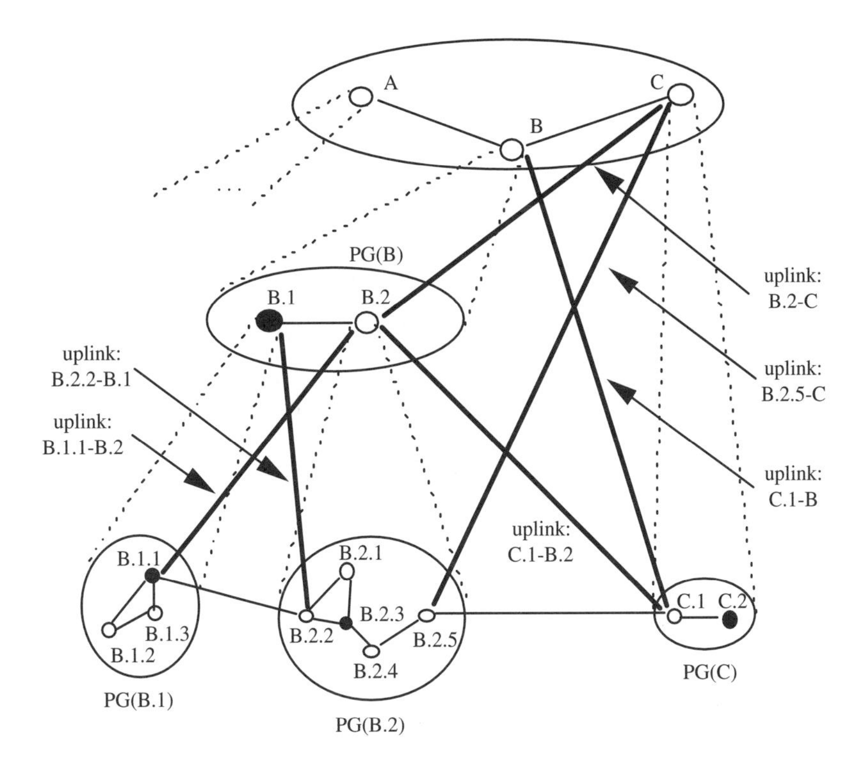

Figure 11.7 Uplinks and upnodes.

The creation of the PNNI hierarchy can be viewed as the recursive generation of peer groups, beginning with the lowest-level nodes and ending with a single top-level peer group encompassing the entire PNNI routing domain. The hierarchical structure is determined by the way in which peer group identifiers are associated with logical group nodes by configuration.

11.2.6 A Node's View of the PNNI Hierarchy

It is instructive to show how a lowest-level node views the entire PNNI hierarchy shown in Figure 11.6. Figure 11.8 shows the view for all the nodes in the lowest-level peer group A.3. The view is the same for all the nodes in A.3, because flooding within the peer group A.3 ensures that the topology databases of all its members are identical.

Let us consider node A.3.3, and assume that the other nodes in peer group A.3 are all active when A.3.3 comes up. A.3.3 will become aware of the links A.3.3–A.3.4 and A.3.3–A.3.2. Through hello packets, A.3.3 will learn about the topology of peer group A.3, and what ATM addresses are reachable through each node. It will also learn about the uplinks A.3.4–A.4, A.3.1–A.2, A.3.2–B and A.3–B, and it will obtain topology and reachability information of the logical group nodes in peer group A. It will also obtain topology and reachability information about the logical group nodes in the higher-level peer group.

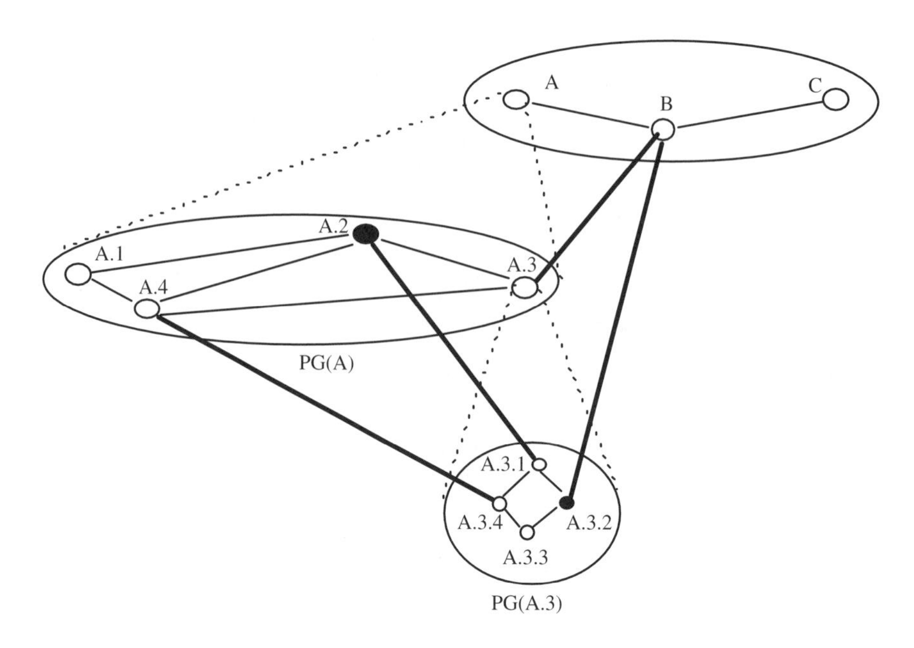

Figure 11.8 A node's view of the PNNI hierarchy.

11.2.7 Address Summarization

Address summarization reduces the amount of addressing information which is needed to be distributed in a PNNI network. It consists of using a single address prefix to represent a collection of end devices and node addresses that begin with the given prefix.

Address prefixes can be either *summary addresses* or *foreign addresses*. A summary address associated with a node is an address prefix that is either explicitly configured at the node, or it takes some default value. A foreign address associated with a node is an address which does not match any of the node's summary addresses. An address that matches one of the node's summary addresses is called a *native address*.

The example given in Figure 11.9 is a subset of the PNNI hierarchy shown in Figure 11.6. There are five end devices attached to A.2.1, five end devices attached to node A.2.2, and three end devices attached to node A.2.3. As before, we use symbolic numbers to represent ATM NSAP addresses. The ATM addresses of the end devices attached to node A.2.1 are: A.2.1.1, A.2.1.2, A.2.1.3, Y.2.1.1 and W.1.1.1. The ATM addresses of the end devices attached to node A.2.2 are: Y.1.1.1, Y.1.1.2, Y.1.1.3, Z.2.1.1 and Z.2.2.2. Finally, the ATM addresses of the end devices attached to A.2.3 are: A.2.3.1, A.2.3.2 and A.2.3.3.

We will use the notation P⟨address⟩ to represent a shorter prefix of an address. That is, P⟨A.2.1⟩, P⟨A.2⟩ and P⟨A⟩ are successive shorter prefixes of the address A.2.1.1. An example of configured summary addresses for each node in peer group A.2 is given in Table 11.1. Other summary addresses could have been chosen, such as P⟨Y.1.1⟩ instead of P⟨Y.1⟩ at node A.2.2, and P⟨W⟩ at node A.2.1. The summary address P⟨A.2⟩ could not have been chosen instead of P⟨A.2.1⟩ or P⟨A.2.3⟩, since a remote node selecting a route would not be able to differentiate between the end devices attached to A.2.3 and

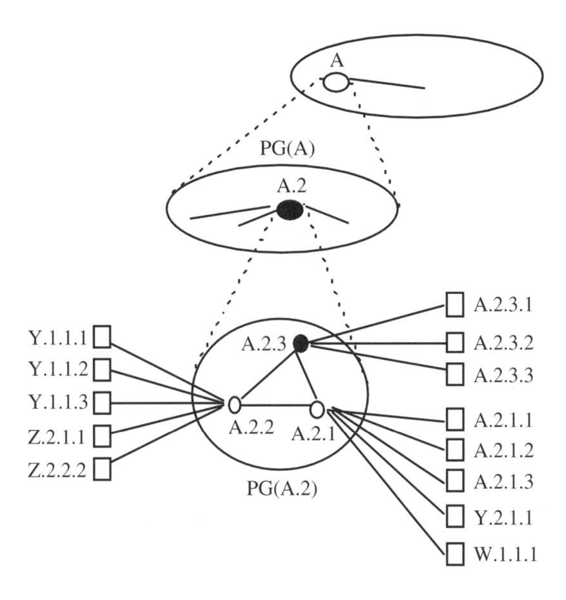

Figure 11.9 End stations attached to nodes A.2.1, A.2.2 and A.2.3.

the end devices attached to A.2.1. We note that the address W.1.1.1 is a foreign address, because it does not match the summary addresses of node A.2.1. On the other hand, the address A.2.3.1 is a native address.

The address prefixes advertised by each node in peer group A.2 are shown in Table 11.2. Node A.2.1 floods its summary addresses plus its foreign address. Nodes A.2.2 and A.2.3 flood summary addresses only since they do not have any foreign addressed end devices.

Table 11.1 Configured summary addresses.

Node	Configured summary addresses
A.2.1	P⟨A.2.1⟩, P⟨Y.2⟩
A.2.2	P⟨Y.1⟩, ⟨Z.2⟩
A.2.3	⟨A.2.3⟩

Table 11.2 Advertised reachable address prefixes.

Node	Reachable address prefixes
A.2.1	P⟨A.2.1⟩, P⟨Y.2, P⟨W.1.1.1
A.2.2	P⟨Y.1, ⟨Z.2
A.2.3	⟨A.2.3

The logical group node A.2 attempts to further summarize the address prefixes flooded in peer group A.2. Specifically, it advertises the following prefixes: P⟨A.2⟩, P⟨Y⟩, P⟨Z.2⟩ and P⟨W.1.1.1⟩.

11.2.8 Level Indicators

As we have seen, PNNI peer groups occur at various levels. Each peer group is associated with a peer group identifier, which is a prefix of ATM NSAP addresses. The length of prefix, that is the number of bits it consists of, is known as the *level indicator*, and it indicates the level of the peer group within the PNNI hierarchy. The level indicator ranges from 0 to 104 bits. An example of level indicators is given in the PNNI hierarchy associated with problem 1 at the end of this chapter (the level indicators are in bold).

PNNI levels are not dense, in the sense that not all levels are used in a specific topology. For example, a peer group with an identifier of length n bits may have a parent group whose identifier ranges anywhere from 0 to $n - 1$ bits in length. Similarly, a peer group with an identifier of length m bits may have a child peer group whose identifier ranges anywhere from $m + 1$ bits to 104 bits in length. Similar level indicators are used for nodes and links.

11.2.9 Path Selection

Due to the connection-oriented nature of ATM, a source end device cannot transmit data to a destination end device unless a connection is first established. We recall from the previous chapter that a source end device requests the establishment of a connection by issuing a SETUP message to its ingress switch. The SETUP message contains a variety of information, such as the ATM address of the destination end device, the amount of traffic the source wants to submit to the network, and the quality-of-service it expects from the network.

The ingress switch is responsible for the establishment of the connection to the destination end device. It first calculates a path to the egress switch of the destination end device. It does that using its local knowledge of the topology of the network that it has acquired through the PNNI routing protocol. Then, it forwards the SETUP message to the next switch on the path, which decides whether to accept the new call or not by running its call admission control algorithm (see Section 7.6). If it accepts the new call, it propagates the SETUP message to the next switch in the path, and so on until the SETUP request reaches the egress switch, which forwards it to the destination end device. If the destination end device accepts the call, then a CONNECT message is propagated back to the ingress switch following the opposite path. The VPI/VCI values are set-up and bound in the switching table of each switch at that time.

There are two basic routing techniques used in networking: *source* routing and *hop-by-hop* routing. In source routing, the ingress switch selects the path to the destination. Other switches on the path simply obey the ingress switch's routing instructions. In hop-by-hop routing, each switch independently selects the next hop for that path, which results in progress towards the destination. Source routing is used in ATM networks, whereas hop-by-hop routing is used in IP networks.

The path that the ingress switch calculates is encoded in a *Designated Transit List* (DTL), which is included as an information element in the SETUP message. The DTL specifies every node used in transit across the peer group of the ingress switch, and it may optionally specify the logical links to be used among the nodes. The path outside its peer group is not specified in detail. Rather, it is abstracted as a sequence of logical group nodes to be transited. When the SETUP message arrives at a logical group node, the node is responsible for selecting a lower-level source route across the peer group that it represents, so that the SETUP message reaches the next hop destination specified in its DTL.

If a node along the path is unable to accept a set-up request, then the node *cranks it back* to a node upstream the path that is allowed to chose an alternative path. Cranckback is used when the path to the destination cannot be found, or when the requested ATM service is not supported by the switch, or when the switch cannot provide the requested bandwidth or the requested quality of service. It is also used when a DTL processing error occurs.

11.3 THE PNNI SIGNALING PROTOCOL

PNNI signaling is used to dynamically establish, maintain and clear ATM connections at the private network-network interface and at the private network node interface. The PNNI signaling protocol is based on the ATM Forum's UNI signaling, and some of its features have been derived from the frame relay NNI signaling defined in ITU-T draft recommendation Q.2934.

The PNNI signaling protocol consists of two distinct entities: *PNNI call control* and *PNNI protocol control*. PNNI call control serves the upper layers for functions such as resource allocation and routing information. The PNNI protocol control entity provides services to the PNNI call control. It processes the incoming and outgoing signaling messages, and it uses SAAL, the signaling AAL, as shown in Figure 11.10. For nonassociate signaling, the signaling channel with VPI = 0, VCI = 5 is used between two nodes. For associate signaling, an available value for VCI is selected within the virtual path connection.

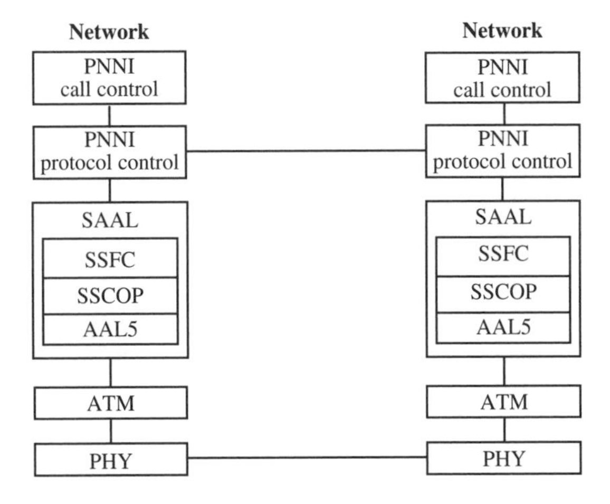

Figure 11.10 The PNNI control plane.

Table 11.3 The PNNI messages for point-to-point call control.

Call establishment messages	ALERTING CALL PROCEEDING CONNECT SETUP
Call clearing messages	RELEASE RELEASE COMPLETE
Miscellaneous messages	NOTIFY STATUS STATUS ENQUIRY

The signaling messages for point-to-point connections are the same as in Q.2931, except the CONNECT ACKNOWLEDGMENT message which is not supported, and they are given in Table 11.3. The signaling messages for point-to-multipoint connections are the same as in Q.2971.

The signaling messages use the information elements described in Section 10.7.1, with some modifications. In addition, the following new information elements were defined: calling party soft PVPC or PVCC, called party soft PVPC or PVCC, crankback, and designated transit list. The first two information elements are used in relation with soft permanent virtual path (soft PVP) or permanent virtual channel (soft PVC) connections. The crankback information element is used to indicate the level of the PNNI hierarchy at which the connection is being cranked back, the logical node identifier at which the connection was blocked, the logical node identifier preceding the blocking logical node on the connection's path, the logical node identifier succeeding the blocking node on the connection's path, and the reason why the call has been blocked. The designated transit list information element contains the logical nodes and logical links that a connection is to traverse through a peer group at some level of hierarchy.

PROBLEMS

1. Consider the PNNI hierarchy shown in Figure 11.11.
   (a) Identify all the border nodes.
   (b) Identify all the logical group nodes (the peer group leaders are marked in black).
   (c) Identify all the uplinks and the induced uplinks.
   (d) What is the view of the PNNI hierarchy from node A.1.2?
   (e) Node A.2.2 receives a request from an end device that is attached to it to set-up a connection to an end device which is attached to node C.1. Describe the path that it will supply in its designated transit list DTL.

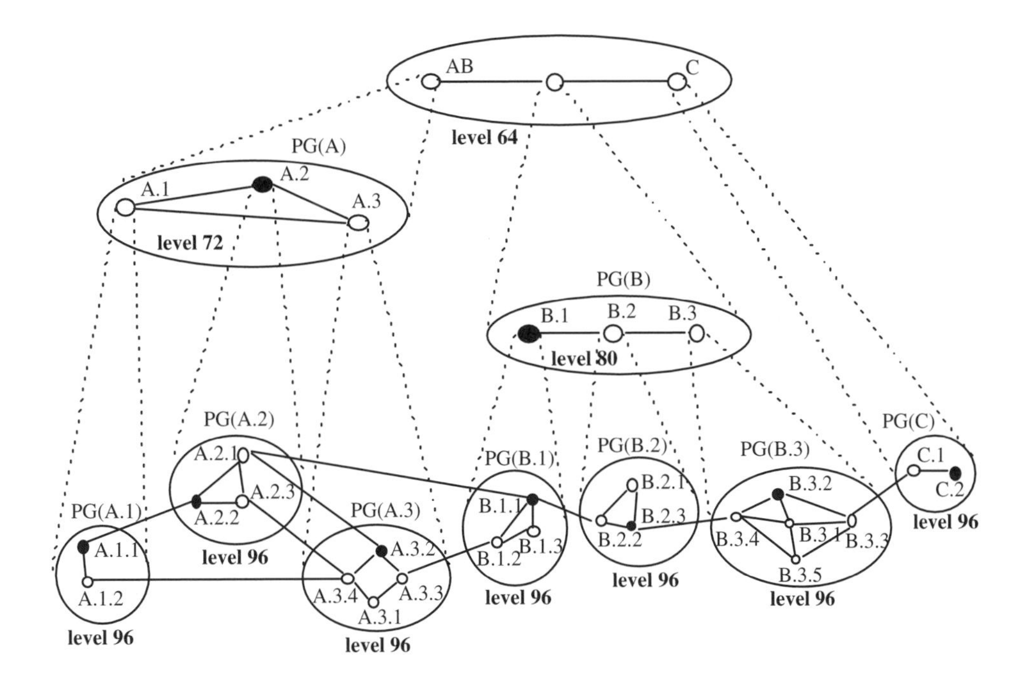
AB
C
level 64
PG(A)
A.1
A.2
A.3
level 72
PG(B)
B.1
B.2
B.3
level 80
PG(A.2)
A.2.1
A.2.2
A.2.3
level 96
PG(A.1)
A.1.1
A.1.2
level 96
PG(A.3)
A.3.1
A.3.2
A.3.3
A.3.4
level 96
PG(B.1)
B.1.1
B.1.2
B.1.3
level 96
PG(B.2)
B.2.1
B.2.2
B.2.3
level 96
PG(B.3)
B.3.1
B.3.2
B.3.3
B.3.4
B.3.5
level 96
PG(C)
C.1
C.2
level 96

Figure 11.11

Appendix: List of Standards

In this section we give a list of some of the standards which are relevant to the topics presented in this book. The standards are listed per chapter.

Chapter 4: Main Features of ATM Networks

B-ISDN General Network Aspects, ITU-T Recommendation I.311, March 1993.
B-ISDN ATM Layer Specification, ITU-T Recommendation I.361, February 1999.

Chapter 5: The ATM Adaptation Layer

Broadband ISDN—ATM Adaptation Layer for Constant Bit Rate Services Functionality and Specification, ANSI, T1/S1 92-605, November 1992.
B-ISDN ATM Adaptation Layer Type 2 Specification, ITU-T Recommendation I.362.2, November 1996.
B-ISDN ATM Adaptation Layer (AAL) Specification, ITU-T Recommendation I.363, March 1993.

Chapter 7: Congestion Control in ATM Network

Traffic Management Specification Version 4.1, ATM Forum, March 1999.
Addendum to Traffic management V4.1 for an Optional Minimum Desired Cell Rate Indication for UBR, ATM Forum, July 2000.

Chapter 8: Transporting IP Traffic Over ATM

LAN Emulation Over ATM Version 1.0, ATM Forum, January 1995.
Multiprotocol Encapsulation Over ATM Adaptation Layer 5, IETF, RFC 2684 (replaces RFC 1483), September 1999.
Classical IP and ARP Over ATM, IETF RFC 2225, April 1998.
Support for Multicast Over UNI 3.0/3.1 based ATM Networks, IETF, RFC 2022, November 1996.
Multicast Server Architectures for MARS-Based ATM multicasting, IETF, RFC 2149, May 1997.
NBMA Next Hop Resolution Protocol (NHRP), IETF, RFC 2332, April 1998.
Cisco Systems' Tag Switching Architecture Overview, IETF, TFC 2105, February 1997.
Multiprotocol Label Switching Architecture, IETF, Internet Draft, August 1999.
LDP Specification, IETF, Internet Draft, August 2000.

Chapter 9: ADSL-Based Access Networks

Data-Over Cable Service Interface Specifications—Radio Frequency Interface Specification, Cable Television Laboratories, 1999.
Broadband Optical Access Systems Based on Passive Optical Networks (PON), ITU-T Recommendation G.983.1, October 1998.
Network and Customer Installation Interfaces—Asymmetrical Digital Subscriber Line (ADSL) Metallic Equipment, ANSI, T1.413 Issue 2, June 1998.
Broadband Service Architecture for Access to Legacy Data Networks Over ADSL Issue 1, ADSL Forum TR-012, June 1998.
ATM Over ADSL Recommendation, ADSL Forum, March 1999.
References and Requirements for CPE Architectures for Data Access Version 3, ADSL Forum WT-31, March 1999.
PPP Over AAL5, IETF RFC 2364, July 1998.
Layer Two Tunneling Protocol "L2TP", IETF, Internet-Draft, November 2000.
Remote Authentication Dial in User Service (RADIUS), IETF, RFC 2865, June 2000.
A Method for Transmitting PPP Over Ethernet (PPPoE), IETF, RFC 2516, February 1999.

Chapter 10: Signaling over the UNI

ATM User-Network Interface (UNI) Signaling Specification, Version 4.0, ATM Forum, July 1996.
Broadband Integrated Services Digital Network (B-ISDN)—Digital Subscriber Signalling System No 2 (DSS 2)—User-Network Interface (UNI) Layer 3 Specification for Basic Call/Connection Control, ITU-T Recommendation Q.2931, 1995.
Broadband Integrated Services Digital Network (B-ISDN)—Digital Subscriber Signalling System No 2 (DSS 2)—User-Network Interface (UNI) Layer 3 specification for Point-to-Multipoint Call/Connection Control, ITU-T Recommendation Q.2971, October 1995.

Chapter 11: The Private Network-Network Interface (PNNI)

Private Network-Network Interface Specification Version 1.0 (PNNI 1.0), ATM Forum, March 1996.

Index